Supreme emergency

Manchester University Press

Supreme emergency
How Britain lives with the Bomb

Andrew Corbett

MANCHESTER UNIVERSITY PRESS

Copyright © Andrew Corbett 2022

The right of Andrew Corbett to be identified as the author of this work has been asserted by them in accordance with the Copyright, Designs and Patents Act 1988.

Published by Manchester University Press
Oxford Road, Manchester M13 9PL

www.manchesteruniversitypress.co.uk

British Library Cataloguing-in-Publication Data
A catalogue record for this book is available from the British Library

ISBN 978 1 5261 4736 3 hardback

First published 2022

The publisher has no responsibility for the persistence or accuracy of URLs for any external or third-party internet websites referred to in this book, and does not guarantee that any content on such websites is, or will remain, accurate or appropriate.

Typeset
by Cheshire Typesetting Ltd, Cuddington, Cheshire

Contents

Acknowledgements	vi
Abbreviations	vii
Introduction: an insider's view	1
1 *The War Game*, a case study	6
2 Government, public and total war (1915–40)	18
3 Government, public and total war (1940–45)	36
4 From the Second World War to continuous at-sea deterrence	55
5 The Polaris replacement decision	98
6 Ethical considerations and wicked issues	144
7 British nuclear deterrence in the 21st century	177
Conclusion: dirty hands and the supreme emergency	224
Bibliography	228
Index	254

Acknowledgements

No book is ever an individual effort – I may have done the writing, but I would not have done much of that without the patient support and guidance of Professor David Whetham and Professor Wyn Bowen, for which I am very grateful. I am very grateful to those senior policy-makers who gave so freely of their time to discuss some deeply visceral issues so openly with me. I hope I have done justice to the depth of feeling and moral engagement which they shared. I'd also like to extend my thanks to Dr Kristan Stoddart for his encouragement and ruthlessly constructive advice on this draft. Obviously the remaining errors and omissions are entirely my fault.

I could not have achieved any of this without endless encouragement from my wife Dr Cynthia Larbey, who helped me persuade myself into doing this project and has supported me throughout, including those fruitless evenings trying to find out why the refences do not match any more; I would like to thank Cynthia for that and so many things.

Abbreviations

ABM	anti-ballistic missile
ACAE	Advisory Committee on Atomic Energy
AEBC	Agriculture and Environment Biotechnology Commission
ANF	Atlantic nuclear force
BBC	British Broadcasting Corporation
BBFC	British Board of Film Censors (now Classification)
BNDSG	British Nuclear Deterrent Study Group
CASD	continuous at-sea deterrence
CENTO	Central Treaty Organisation
CND	Campaign for Nuclear Disarmament
CO	commanding officer
COS	Chiefs of Staff
DAC	Direct Action Committee
DHSS	Department of Health and Social Security (since 2001, responsibility largely under Department of Work and Pensions)
DOPC	Defence Overseas Policy Committee
DS19	(MOD) Defence Secretariat 19
DSMA	Defence and Security Media Advisory Committee
DUS(P)	deputy under-secretary (policy)
FCO	Foreign and Commonwealth Office
GLCM	ground-launched cruise missile
HC Deb.	House of Commons Debates
HL Deb.	House of Lords Debates
HLG	(NATO) High Level Group
INF	intermediate-range nuclear forces

IRBM	intermediate-range ballistic missile
IVF	*in vitro* fertilisation
LRTNF	long-range tactical nuclear forces
MLF	(NATO) Multi-Lateral Force
MOD	Ministry of Defence
MRBM	medium-range ballistic missile
MRC	Medical Research Council
NATO	North Atlantic Treaty Organisation
NCAWT	National Council for the Abolition of Nuclear Weapons Tests
NPG	(NATO) Nuclear Planning Group
NPT	Nuclear Non-Proliferation Treaty
OR	(Air Staff) Operational Requirement
RAF	Royal Air Force
RUSI	Royal United Services Institute
SALT	Strategic Arms Limitation Talks
SDSR	Strategic Defence and Security Review
SIOP	single integrated operating plan
SLBM	submarine-launched ballistic missile
SLCM	submarine-launched cruise missile
SSBN	ship – submersible, ballistic, nuclear
TNA	The National Archives
UNSC	United Nations Security Council
USSR	Union of Soviet Socialist Republics
WAC	BBC Written Archive Centre

Introduction
An insider's view

I am not an ethicist, but I have given a great deal of thought to the morality of the use of force and, in particular, the concept of nuclear deterrence. I have served in Polaris and Trident ballistic-missile submarines (ship – submersible, ballistic, nuclear or SSBN) on and off since 1986, including command of two Vanguard Class submarines, HMS *Vengeance* and HMS *Vanguard*, between 2003 and 2007. I therefore have had ample opportunity, and motivation, to reconcile the full potential of my personal responsibilities with some kind of moral compass.

I was always content with the ethical and strategic aspects of my responsibilities, and I would discuss them with my ship's company, but I could never be sure that I could articulate them in terms of which the Ministry of Defence (MOD) would approve. So, I asked within my 'command chain'; fruitlessly. In hindsight, frustration at the inability of the MOD to provide official guidance to the commanding officer of an SSBN ultimately led me here; if the MOD cannot articulate the official rationale for the UK's Strategic Nuclear Deterrent to its own SSBN commanding officers, what does that say for its ability to make a coherent argument in public? In the absence of formal guidance, I considered the available contemporary academic literature. The rather binary options available seemed initially to be that either all those involved in the business of deterrence were acting immorally, or there was a shortfall in the literature. I do not accept that all of the very honourable professionals with whom I served are merely immoral, or too stupid to notice, or too hypocritical to care. Nor are they amoral in the Machiavellian realist sense; neither I, nor they, accept that there are no appropriate rules that should be governing this highly emotive and very difficult area of moral thinking.

This book is my contribution to evolving a better understanding of those rules, and why successive British governments have struggled to articulate nuclear deterrence policy, both internally and publicly. My core hypothesis is that there are always moral principles that apply to the use of force, but in situations of extreme peril – a supreme emergency – the consequences of failure may be so horrendous that action to avert it might exceptionally breach usually accepted norms of behaviour. For those involved in such decisions, moral discussion of those circumstances is complex enough *in camera*, has proven extremely demanding in Parliament and is increasingly becoming well-nigh impossible in modern media.

It is important at this stage to consider the concepts of 'the public' and 'the media'. There is a great deal of very powerful analysis of both of these concepts, but I do not intend to address it here. The issues I do address tend to focus on how 'the public' and 'the media' were perceived by government and influenced government decision-making at particular times, so I have used a very reductionist view of both terms; 'the public' are those members of the British electorate who could vote in general elections. This includes active opponents of nuclear policy. 'The media' comprise those organisations that publish in print, radio, television or other digital means in order to inform or influence policy and 'the public'. I have not referred to public opinion at all, other than when it is considered by government, because it too is a very nebulous concept, which of itself is not pertinent to the study I make here, other than when the evidence suggests that the government perception of public opinion specifically influenced government thinking.

This argument will proceed in five stages. Chapter one will consider the case of *The War Game*, a BBC documentary-drama that was made in 1965 but not shown on British TV until 1985. It illustrates clearly the close linkage between political perceptions of public understanding and opinions, the reticence of government (in its broadest sense) when it comes to public consideration of nuclear deterrence and the distortion that this reticence imposes on government presentation of policy.

The second stage (chapters two and three) is a historical retrospective which considers the moral debates associated with aerial bombardment in the First and Second World Wars. This complex morality was well understood by Prime Minister Churchill and his Cabinet, and the senior members of the armed forces who directed operations against the Nazi regime. Churchill advocated abrogation of neutral states' rights in a note to Cabinet describing the concept of the 'supreme emergency' in December 1939:

> Our defeat would mean an age of barbaric violence ... we have a right, indeed are bound in duty, to abrogate for a space some of the conventions of the very laws we seek to consolidate and reaffirm ... The letter of the law must not in supreme emergency obstruct those who are charged with its protection and enforcement ... Humanity, rather than legality, must be our guide.[1]

This view of a 'supreme emergency' and its relationship with moral constraint form the core thread of my argument.

The British government reacted with moral outrage to the German bombing of undefended British cities during the First World War (chapter two), but participated in the protracted and still hotly contested strategic bombing campaign of the Second World War which caused tens of thousands of civilian casualties in German cities. There were intense discussions in Cabinet, in the Air Staff and in Parliament about the moral implications of this campaign (chapter three). I do not seek to reach a definitive position on the morality of that campaign, but do endeavour to show that the British government's public position on the capability of and intent behind the bombing campaign, certainly from 1943, bore little relation to its understanding of the strategic realities; and I argue that this was because of the difficulties of making a complex moral case in public.

The third stage (chapters four and five) charts the impact of this reticence as the UK developed a nuclear deterrent capability, and its impact on the excessive secrecy surrounding every aspect of British nuclear deterrence policy from the decision to develop a sovereign capability in 1947, via the decision to replace Polaris in 1980 and the protracted decision-making process to replace the Vanguard class submarines in the early 21st century.

The last two chapters of the book are more about the issues involved. The ethics of nuclear weapons policy are considered, both as deterrence policy evolved, and in the strategic environment in 2020. This leads to a consideration of the moral aspects of 'wicked problems': those complex decisions which have no simply good or bad outcomes, and the issues that this complexity poses for public policy. A short case study of government handling of another, equally complex, moral issue is considered in parallel. Intense public debate about *in vitro* fertilisation (IVF) between 1978 and 1982 led to the establishment of the Warnock Enquiry into Human Fertilisation and Embryology to investigate the social, ethical and legal implications of IVF. The Enquiry's report was published in 1984 and led to the Human Fertilisation and Embryology Act of 1991 – to no particular public interest or media hysteria. Yet it is clear that, despite the increased openness of the Thatcher years and the Open Government Documents of the 1980s, the government remained on the back foot with the public discourse on nuclear weapons policy.

The final chapter of the book will address 21st-century deterrence, and seek to place nuclear deterrence in its current strategic context. This highlights the linkage between ethics, culture, technology, domestic and international factors in the formulation of national security policy and the role of nuclear weapons within it, and the public presentation of that policy and strategy.

In short, this book is offered as a one-stop shop for the reader interested in nuclear deterrence in Britain, its derivation and role in strategy, and the broader interaction between the public and the government in formulating this cornerstone of British security policy. There are many very good, more detailed examinations of each of these factors, and I indicate suggestions where the reader might want to pursue more detail. In the end, though, this is a personal account; it portrays my understanding of why and how the UK procured and sustained a strategic nuclear deterrent, the moral issues involved, their relations with strategic decisions, the impact they had on the political decision-makers and the way successive governments felt able to portray these issues to the electorate.

Note

1 Sir Winston Churchill, *The gathering storm*. New York: Rosetta Stone, 2009: 492.

1

The War Game, a case study

The fate of *The War Game*, a radical film about the effects of nuclear weapons, provides a clear illustration of the ambiguity of government engagement with the public on nuclear deterrence policy in the context of the height of the Cold War and puts into context the inquiry that the rest of this book seeks to address.

In May 1965, a fictional BBC television documentary-drama depicting the possible aftermath of a nuclear attack on Britain was completed and the initial draft shown to the Controller of BBC2, Huw Wheldon. Five months later, the BBC announced it had decided not to air the programme on television because it was an 'artistic failure'.[1] It was subsequently given a limited cinematic release and won the 1966 Venice Film Festival Award for best documentary, and the best documentary Academy Award (Oscar) in 1967. Eventually, the BBC screened it on television on 31 July 1985, twenty years after its completion. To this day, its director insists that the film was suppressed because of its political impact.[2]

From the beginning Wheldon was well aware of the political impact that the film might have and insisted on close supervision of the filming process. The film's director, Peter Watkins, had previously directed *Culloden*, a dramatic depiction of the defeat of the 1745 Jacobite rebels, and his film had been very well received critically and publicly. Watkins had a list of difficult topics which he wanted to address with his docu-drama format, both challenging received wisdom on the subject matter and stretching the use of television as a medium for entertainment and information. Watkins's intent was to 'challenge viewers' assumptions and provoke different perspectives'.[3] Top of his list was nuclear war.

The War Game, *a case study*

Watkins's research was meticulous, including interviews with survivors of Second World War mass bombing raids, both British and German, and members of the emergency and police services. He also sent a number of detailed questionnaires to government departments and local government offices regionally, inquiring specifically into the detailed preparations being made in civil defence during the early 1960s. From its very inception, *The War Game* was potentially contentious, politically; it set out to depict on television the aftermath of nuclear war in a way that directly challenged the official view.

Inherently pacifist, Watkins was perturbed by the lack of public information and understanding of the nature of nuclear war, and by the government claims about the effectiveness of preparations being made for civil defence within the United Kingdom. This is the central refrain of the film and is literally echoed by the narrator at the end of the film as the camera pans across a group of children orphaned by the nuclear attack:

> On almost the entire subject of thermo-nuclear weapons; problems of their possession; effects of their use; there is now practically total silence; in the press, in official publications and on television. There is hope in any unresolved and unpredictable situation but is there a real hope to be found in this silence?[4]

The War Game is filmed using handheld cameras in a highly dynamic 'newsreel' style, closely reflecting the contemporary footage being returned from the conflict in Vietnam. The cast are almost all amateurs, locals of the town where the film was made. The film portrays plausible outcomes of a nuclear war, starting with the immediate aftermath of blast and heat; already well understood. The memories of the bombing raids of the Second World War were still vivid to many. *The War Game* specifically parodies the civil defence information films being produced by the Home Office; the narrator repeatedly uses the phrase 'This is what nuclear war means', which had been a central motif of *Doom town*, a 1955 Pathé civil defence training film depicting search and rescue in burned-out buildings.[5]

The government Civil Defence Corps had produced a number of training films and publicity 'shorts' in the 1950s and early 1960s.[6]

These films depicted civil defence exercises and scenarios which were scripted to run up to the evacuation of casualties to conveniently located first aid posts where assistance from unaffected areas was available. The rescue services always appeared in control and there was a clear message that civil defence was a viable response to a nuclear attack. *The War Game* scenario develops beyond this point and the film portrays the failure of the Civil Defence Corps to respond to the demands of ever-increasing casualties, including the inadequate provision of medical care, mercy killings of very seriously burned casualties by the police and mass cremations in order to prevent the spread of disease. As time passes in the film, the situation changes from one of immediate emergency to one of protracted crisis attributable to failure of the civil defence organisation, leading to food shortages, looting and finally the imposition of martial law on the streets; graphically depicted in the summary execution of food rioters by police firing squad.

Throughout, this fabricated newsreel footage is interspersed with interviews to camera of equally fictitious 'establishment figures' such as a bishop, government ministers, officials and senior military officers. These interviews, however, are filmed at desks or in offices and they rehearse genuine government statements on civil defence planning; assurances that procedures and processes are in place to ensure that nuclear war is survivable. Immediately after each official statement, the film depicts its 'reality', directly in counterpoint to the reassuring view of the 'establishment figure'. This reality includes live unscripted interviews with members of the public who were actually participating in the filming. They were asked real questions and answered from their real viewpoints. Watkins later said 'And those questions and responses – particularly the responses – are perhaps the biggest single indictment in the entire film of the way we are conducting our present society and of the lack of common public knowledge of the things which affect humanity'.[7] Ultimately, the effect is that the fictional newsreel and interview footage completely discredits the genuine government statements and Civil Defence Organisation assurances.

The decision not to show *The War Game*

Even before filming started Wheldon had been in close liaison with Grace Wyndham Goldie (Head of Talks at the BBC) who supported the film in principle: 'so long as there is no security risk and the facts are authentic, the people should be trusted with the truth ...'[8] This is the essential question here; assuming Watkins's film was authentic, what was it that prevented government trusting people with the truth?

In parallel, there was an ongoing dialogue with the Home Office throughout filming, with Wheldon insisting on editorial independence despite:

> ... the Home Office argu[ing] that as 'partners in the civil defence field', the government and Corporation ought to work together throughout production to ensure that the film was 'prepared with the utmost care and responsibility' given its potentially harmful effects on the public.[9]

Despite Wheldon's support, *The War Game* caused considerable unease within the BBC hierarchy during its production. Filming was completed in April 1965, and Watkins completed his initial editing by mid-June. The first cut was screened to Watkins's panel of expert consultants on 17 June and to Wheldon and Richard Cawston (Wheldon's replacement as Head of Documentaries) on 24 June. After each of these screenings, the film was edited further. After a further screening to BBC publicity officers, the re-edited film was viewed again by Cawston on 18 August, who gave the film a provisional broadcast date of 7 October, to be followed by *Tonight*. However, on 2 September, Hugh Carleton-Greene (BBC Director-General) and Lord Normanbrook (Chair of the BBC Board of Governors) viewed the film and decided to 'take soundings' from Whitehall. As an ex-Cabinet Secretary, and chair of the 1954 'Committee on Nuclear Defence and Civil Defence', Normanbrook would have been very aware of the potential domestic impact of the film. He wrote to the Cabinet Secretary that the film:

> ... has been made with considerable restraint. But the subject is, necessarily, alarming; and the showing of the film on television might

well have a significant effect on public attitudes towards the ... I doubt that the BBC alone should take the responsibility of deciding whether this film should be shown ...'[10]

When informed of this decision by Wheldon, Watkins resigned over what he saw as political interference in the independence of the BBC.[11]

On 24 September, *The War Game* was screened to senior government officials including the Cabinet Secretary Sir Burke Trend, the Permanent Secretaries from the Home Office and the Ministry of Defence (MOD), and a senior officer representing the Chief of Defence Staff. Trend clearly understood the government position very well; he wrote in October to the Lord President and Prime Minister that *The War Game* was unbalanced and pessimistic about civil defence, but that 'the dilemma for the government ... was that it could not afford to give the impression that, by overriding the BBC's duty to educate, it was sweeping under the rug an issue which ministers found politically embarrassing'.[12] Trend met Normanbrook again on 5 November and Normanbrook noted that the '... decision should be left to the discretion of the BBC ... it is also clear that Whitehall will be relieved if the BBC chooses not to show it'.[13] On 24 November, Normanbrook wrote to Trend to inform him that the BBC had decided not to show the film. A BBC press release on 26 November read 'this is the BBC's own decision ... not as a result of outside pressure of any kind'.[14] *The Times* reported this release, adding '... the film has been judged by the BBC to be too horrifying for the medium of broadcasting'.[15]

There were questions asked about the degree of government involvement in the BBC's decision in both Houses of Parliament, but government officials and ministers, including Prime Minister Harold Wilson, simply reiterated that '[a]s regards rumours about *The War Game*, the Government have not interfered at all'.[16] Peter Watkins remains in absolutely no doubt that the government applied pressure to the BBC to inhibit the showing of *The War Game*.[17]

There was undoubtedly contact between the hierarchy of the BBC and very senior government officials prior to the decision being made. The extent to which that contact manifested as pressure on the BBC is unlikely ever to be established definitively from the official records. There is clearly a degree of historiographical debate about the exact decision-making process that resulted in the censorship of *The War Game*; at least three historians have independently assessed substantially the same evidence and reached substantively different conclusions. Chapman concludes '[t]hat there was pressure from Whitehall (not from Westminster) not to show the film cannot be doubted ...'[18]

In a form letter sent to those who had written to the BBC to complain, the BBC wrote 'There was an element of experiment in this project ... Such programme experiments sometimes fail ...'[19] As shown above, this is not quite the view expressed by Lord Normanbrook; he had admitted on first seeing it that the film was an impressive documentary but he clearly had reservations about its political impact since he also insisted that the responsibility for showing it was too great for the BBC to shoulder alone. 'In making this decision, the BBC had to set aside its own belief that the probable effects of nuclear warfare should be made known to the public, if at all possible, through the medium of television.'[20] This does appear to concur with Wyndham Goldie's comment that the people should be trusted with the truth, but it then rather begs the question why the BBC had to set aside its own beliefs and not show the film. The BBC's letter continued 'most of those who saw it were very deeply affected, and believed that it had the power to produce unpredictable emotions and moral difficulties whose resolution called for balance of judgement of the highest order'.[21]

The War Game vividly challenged the sterile depictions of the aftermath of nuclear war shown in the Civil Defence Corps training films and the Pathé news reports. The opportunity to inform 'balanced judgement of the highest order' is exactly what Watkins was seeking to promote; a genuine debate about nuclear weapons policy. By not showing the film, the BBC did not provide that opportunity, and according to some commentators, deliberately stifled it: '... *The War Game* was censored for politically motivated reasons ... the state was intimately involved in the BBC's decision and that there was nothing "ad hoc" about the process'.[22] The BBC had written in 1965: 'The BBC has, therefore, reluctantly decided that, because of its nature, this film cannot be broadcast ... In making this decision, the BBC acted on its own judgement. There was no outside pressure. In particular, we received no advice from Government Departments or officials about whether or not the film should be shown on the air.'[23] Shaw concludes that '[a]vailable records fail to make it clear precisely who took the decision to pull *The War Game* and when'.[24] Wayne's article (quoted above) was a direct challenge to Chapman, but from their exchange the most salient point for this study is '[i]n its own way, Wayne's passionate and polemical response is further evidence of the controversy that continues to surround the BBC's decision not to broadcast *The War Game* ...'[25]

The BBC consulted government officials, who, whether or not they actually coerced the BBC decision, clearly felt that *The War Game* was in some way 'hazardous'. The Cabinet Secretary's memo to Wilson suggests that the Cabinet Office felt that the government was in a cleft stick; it could neither suppress the film (for fear of being accused of doing so), nor allow it to be aired (for fear of the concerns it would raise).[26]

To date, the analysis of the events surrounding the suppression of *The War Game* has focused on the extent to which the government influenced the BBC decision. The debate quoted above concentrated on whether the government effectively compelled the BBC's decision (Wayne) or was merely a more passive party to it (Chapman and to a lesser extent Shaw). There is, however, a clear consensus that the film was suppressed. But none of these articles addressed in any depth the assumption which underlies all three: why would the government want to suppress *The War Game*?

Much of the information associated with nuclear deterrence capability and limitations is necessarily very highly classified and therefore not at all appropriate for the public domain. In the middle of a war, even a 'cold' war, some information is just too sensitive for public dissemination. In 1962 five members of the radical anti-nuclear protest group the Committee of 100, who had protested outside nuclear and civil defence bases and thus compromised the bases' locations, had been convicted of offences against the Official Secrets Act and had received custodial sentences.[27]

As a result of a Cabinet Office study of 1955 (the Strath Report), the government knew all too well that civil defence procedures and capabilities were inadequate (this will be considered later in this book). *The War Game* promised to ridicule the claims of the Civil Defence Organisation and to challenge government claims that a nuclear war would be survivable. In particular, the loss of control by the Civil Defence Organisation and police forces reflected accurately the concept of 'breakdown' highlighted by the Strath Report and subsequently in very highly classified government studies: 'when the government of a country is no longer able to ensure that its orders are carried out. This state of affairs could come about through breakdown of the machinery of control.'[28] Therefore it could have been argued that there was a national security issue raised by *The War Game* and the government did not want shortfalls to become public (or Soviet) knowledge; hence the desire to have it suppressed. But had security really been a salient issue, the government would have been able to stop transmission of *The War Game* through well-established 'D-notice' procedures, with none of the attendant publicity. There was no concern about security expressed in the government papers of the time; merely that *The War Game* exposed issues that might be 'embarrassing'.

There was an ongoing debate within government as to whether spending on civil defence should be increased or whether to rely on investment in active elements of deterrence on the assumption that war could be deterred in the first place, and the need for civil defence obviated.[29] In general, government decisions about nuclear weapons policy had been kept within very limited circles since Churchill had authorised the Tube Alloys project in absolute secrecy in 1942: 'Mr Churchill had vigorously insisted that knowledge of the atomic bomb be kept to the smallest possible circle of Ministers and advisers'.[30] Veteran parliamentarian Tony Benn notes that, as a junior MP in 1950, 'I tried to put down a question about nuclear weapons, having discovered that the Labour government had built the atomic bomb without telling parliament. I was sternly rebuked by Attlee, which at the time was quite frightening.'[31]

During his second term as Prime Minister, Churchill was more open with Cabinet and Parliament about nuclear matters than Attlee had been, but even so there had been very limited public discourse on nuclear policy in the UK during the 1950s and 1960s.

Even in the face of the Campaign for Nuclear Disarmament's activities, which reached their peak at the turn of the decade, there was persistent government reticence about nuclear matters.

The fate of *The War Game* acts as a useful case study into this reticence. In February 1966, four months after the decision not to air *The War Game*, the BBC arranged a cinematic screening to an invited audience of MPs, journalists (mostly war correspondents) and others. It received reviews ranging from 'a warning masterpiece; It may be the most important film ever made'[32] to 'muddle-minded Mr Watkins'[33] and 'one ban the BBC need not have defended'.[34] In March 1966, the British Film Institute was given a limited licence to screen *The War Game* and it was granted an X certificate by the British Board of Film Censors: '... *The War Game* was a brilliant film which he [the secretary of the BBFC] thought should be shown in cinemas'.[35] On the basis of this cinematic release, *The War Game* was awarded 'Best Documentary' at the 1967 Academy Awards; 'Best Short Film' at the 1967 British Academy Film and Television Awards (BAFTA); it won the 1967 Bilbao International Festival of Documentary and Short Films and Watkins was awarded special prizes by BAFTA (1967) and the Venice Film Festival (1966).

The film itself was judged *in absentia* by newspapers and the fate of the film became the story, rather than the story being about the scenario it depicted and the issues it raised. Shortly before the licence was granted to the BFI Mrs Stella Reading, the Chair of the Women's Voluntary Service for Civil Defence, wrote:

> The reported decision of the BBC to make *The War Game* available to the National Film Institute for public showing ... is to my mind a grave error of judgement ... If the argument is that the film should be shown in the interests of truth, then let us have *all* the truth, not only the certainty of enormous destruction and loss of life but the equal certainty that with efficient national preparation many millions of lives could be saved.[36]

By 1966, the (top secret) Strath Report was a decade old. Nuclear weapons had grown far more destructive and the damage that could be expected in the UK in the late 1960s was exponentially greater than that in 1955. However, in line with the decision to rely on deterrence, civil defence funding had been reduced, not increased, after 1960 and Civil Defence Corps numbers had fallen from a 1960 peak of 360,000 to 140,000 in 1965. In 1965, the Corps was put into a minimal budget 'care and maintenance' stance by the Labour Government Home Defence Review.[37] It is not clear whether Mrs Reading knew of the Strath Report, in which case her letter would have been a piece of breathtaking hypocrisy, or (far more likely) was completely ignorant of the Report and its ramifications, despite her position of apparent authority, and was simply convinced of the efficacy of 'efficient national preparation'.

The BBC made the decision not to show *The War Game*, and then made the decision to allow it a limited cinematic release. It was an Oscar-winning success, not an experimental failure. It depicted a narrative of nuclear war that was familiar to government officials but not the public. The concept of breakdown was officially understood yet 'Civil Defence Handbook No 10' (the version current in 1965)[38] and the civil defence training and information films were sanitised to the point of sterility, particularly on the prospects for law and order following an attack. The arguments about the accuracy of the fiction depicted by *The War Game* were played out in public, but by proxy.

Whether or not the BBC asserted the independence of its decision not to show *The War Game*, it did decide only after extensive and detailed consultation with government officials, which both sides denied publicly. The debate in the letters sections of the press was conducted by the Chair of the Women's Voluntary Service for Civil Defence and other quasi-government spokespeople, or especially interested Members of Parliament, from either camp. Anti-nuclear lobbyists were vocal, the film gave the waning membership of the Campaign for Nuclear Disarmament (CND) a rallying point, but this was short-lived. The one participant in this debate conspicuous by its absence is an authoritative government voice; why?

In 1965, there was almost total silence on the subject of nuclear weapons. The withdrawal of *The War Game* highlighted that issue but even then, the government did not intervene in discussion either about the fate of the film, or about the core issue. This book considers what factors led to the 'total silence' to which the *The War Game*'s narrator called attention.

Notes

1 Peter Watkins, personal website: http://pwatkins.mnsi.net/index.htm [accessed 3 January 2013].
2 Peter Watkins, interview with A. S. Corbett, 27 September 2012.
3 Watkins, interview with Corbett.
4 *The War Game*, Peter Watkins (1965), London: BBC.
5 *Doom town* [Newsreel] (1955), London: Pathé.
6 *Nuclear war in Britain; home front civil defence films 1951–1987* (2010), Strikeforce TV.
7 Alan Rosenthal. *New documentary in action: casebook in film making* (Berkeley, CA: University of California Press, 1971), 160.
8 BBC Written Archive Centre (WAC) T56/263/1: Wheldon to Kenneth Adam, 31 December 1964.
9 BBC WAC T16/679/1: Winther to Arkell, 22 January 1965; R 78/2, 680/1: Winther to Arkell, 16 February 1965; Arkell to Carleton Greene, 18 February 1965.
10 BBC WAC, T16/679/1: Controllers' Minutes, 6 September 1965.
11 Watkins, interview with Corbett.
12 Trend to Wilson, 8 October 65; S251 box 8 'The Story of *The War Game*', both in The National Archives (TNA) PREM 13/139.
13 BBC WAC, T16/679/1: Normanbrook to Carleton Greene, 5 November 1965.
14 BBC WAC, R44/1334/1 F. L. Cobb (acting day press officer), 26 November 1965.
15 'Film "too horrifying" for television', *The Times*, 27 November 1965, p. 4.
16 Prime Minister Harold Wilson, 30 November 1965, *Hansard*, 5th series, vol. 721, col. 1230.
17 Watkins, interview with Corbett.
18 James Chapman, 'The BBC and the censorship of *The War Game* (1965)', *Journal of Contemporary History* 41 (2006), 75–94: 93.
19 BBC unreferenced letter signed 'O. G. TAYLOR' dated 23 December 1965. Peter Watkins's private archive.
20 *Ibid.*
21 *Ibid.*
22 Mike Wayne, 'Failing the public: the BBC, *The War Game* and revisionist history: a reply to James Chapman', *Journal of Contemporary History* 42 (2007), 11: 627.
23 *Ibid.*

been a source of ethical debate, prompting an odd blend of public pugnacity, political timidity and distaste in Britain, since the First World War. The very tenor of both strategic thinking and the laws of war were challenged by the experiences of the Great War and its emerging technologies, threatening non-combatants in a way inconceivable to earlier strategists and ethicists. The refinement of these technologies between the wars led to an almost tragic inevitability about the fate of Hiroshima.

This chapter will investigate the context in which the decisions were made that led to the British bombing campaign against Germany in 1943–5, not the legitimacy of the campaign *per se*. It will address the evolution of the technical, operational, doctrinal, political and cultural factors that combined to influence those decisions, identify the most salient factors and examine how those decisions and factors were presented to the public.

Initial examination of the factors that influenced decisions made about strategic bombing during the Second World War suggests that many had their roots in the earliest experiences of aerial bombardment during the First World War. The reactions of British politicians, military planners and the public to the bombing of civilians by German aircraft during the First World War created precedents for much of the debate to follow in the 1940s.

The First World War

British politicians and newspapers responded to Zeppelin attacks on London in the autumn of 1915 with relative restraint; one headline read 'Zeppelin raid on London; eight persons killed and over thirty injured'.[3] The general theme of the public and press reaction was one of regretful bemusement at this apparent breach of the rules of 'civilised warfare'. *The Guardian*, quoting an American interview with a German airship commander who asserted that he felt it deeply when '... he learns that women and children and other non-combatants are killed' concluded that '... one of the blackest of the many crimes with which Germany has stained herself during this past year is that she has introduced this inevitably haphazard murder into warfare'.[4]

Debates were held in the Commons over the precautions being taken against German air raids, but the government consistently refused to discuss details of capabilities, tactics or policy on grounds such as '[i]nformation as to the numbers, disposition, and efficiency of the guns available for anti-aircraft defence would be exceedingly valuable to the enemy, and cannot therefore be made public'.[5] In what was to become characteristic of discussion of strategic bombing, the first government response was to plead secrecy. After a six-week campaign including twenty-six raids against the capital, but with increasing fatality rates amongst the airship crews, the Zeppelin raids were stopped; 550 civilians had been killed and more than 1,300 injured.[6]

On the evening of 20 March 1917, the hospital ship *Asturias* was torpedoed off Devon by the submarine UC-66 with the loss of thirty-five lives. Since it was almost impossible to defend such ships, the Cabinet discussed the issue of reprisals on 3 April but deferred decision on this difficult issue until more definitive information about the attacks was known.[7] A week later they agreed 'that the only practicable form of reprisals ... was the aerial bombardment of an open German town ... though most reluctant to embark upon a policy which might involve the killing of women and children ... there was no alternative'.[8]

In Parliament, the issue was more black and white; reprisals for attacks on hospital ships and aerial bombardment of towns were either expedient or abhorrent. Sir John Lonsdale demanded '... what steps are to be taken to give effect to the threat of reprisals against Germany for torpedoing hospital ships?'[9] Throughout the summer, Mr Pemberton Billing argued vehemently for reprisals against the renewed air raids: '[t]here is only one way by which we shall be able to stop air raids of heavier-than-air machines; that is by reprisals. Whether this country likes reprisals or not bothers me very little.'[10] Others such as Mr Molteno were equally opposed: 'does the War Cabinet think that it would be in accordance with the high principles of humanity if fighting men are withdrawn from fighting the armed forces of the enemy to attack the civil population?'[11]

Eight bomber aircraft of No. 2 Squadron of the Royal Naval Air Service twice attacked the German town of Freiburg on 14 April,

Government, public and total war (1915–40)

dropping about thirty 65lb bombs and leaflets explaining why the town was being attacked.[12] The newspapers reported the attack specifically as a reprisal, the *Guardian* headline reading 'Hospital ship crimes; naval airmen bomb German town'.[13] There was no comment on the British adoption of the same 'form of 'haphazard murder' deplored in 1915.

The first German Gotha aircraft bombing attacks on Dover and Folkestone in May 1917 drew an emotive press response: 'Shrapnel for women and children'.[14] This *Guardian* editorial continued '[i]f we thought that reprisals on undefended towns would be effectual in stopping German raids on our own, we might, for the sake of saving life, have to waste it, repulsive as the necessity might be'. A subsequent attack on London on 13 June hit a school, killing eighteen children. The debate in the Commons the next day contained no rhetoric, nor debate about retaliatory attacks; it concentrated on the defences for London (which the government still refused to describe)[15] and on the matter of air raid warnings: 'I am informed that yesterday people in the West End actually took taxis to go down to see the raid when they heard that it was taking place'.[16] The Cabinet noted that 'out of all the casualties caused by the raid there was not the name of a single soldier'.[17]

Whilst the famously succinct Cabinet minutes seldom indicate the full flavour of the discussions, the note here that only non-combatants were killed would suggest that the issue of the targeting of non-combatants had been significant in those discussions. Cabinet agreed to discuss 'reprisals for air raids, with a view to the investigation more particularly of the effect which the adoption of a policy of reprisals would have on the aerial operations on the Western Front'.[18] The press reporting was becoming more critical, both of the German action and the British response. *The Guardian* headline read 'Many bombs in East End; school hit. Ten children killed.'[19] *The Times* editorial demanded more defences for London, saying '[w]e have constantly pointed out that aeroplane raids of this character will become larger and more frequent'.[20] *The Guardian* editorial also demanded better defences but concluded '[i]t may be taken for granted that the Germans will repeat yesterday's raid on London, if possible with exaggeration of its horrors, as often as they dare ... moral considerations are expressly ruled out of account in the calculations of the German General Staff'.[21] These appear to be early steps in a trend by the British press to assume the moral vacuity of hostile nations. There were, however, more radical recriminations, with some public meetings demanding reprisals; *The Times* reported one meeting attended by thousands which passed a resolution calling for air reprisals 'amid great cheering'.[22]

The German air raids continued, with dissent growing over the apparent impunity of the attackers. In late June the Cabinet Secretary received apparently coordinated appeals from residents' groups calling 'on the Government to initiate immediately a policy of ceaseless air attacks on German towns and cities, in order that their populations may experience the effect of their own methods of warfare, and thus [be] induced to force the German authorities to cease this wanton destruction of life and property'.[23]

Twenty-six civilians were killed on the morning of 7 July and that afternoon Cabinet decided that 'immediate steps must be taken to prevent a recurrence of these raids. Two methods of dealing with this question were discussed ... The maintenance of an efficient force of machines in England with which to repel attacks [and] ... Counter-attacks to be made on German towns'.[24] Two squadrons of aircraft were withdrawn from the 'front' and deployed to enhance the defences of south-east England. No reprisal action was ordered.

Cabinet was dealing with a complex problem. There was a clear imperative to do something, but defences were ineffective and there was clearly a desire not to retaliate against non-combatants. The majority of the Cabinet were also of the opinion that reprisals should only be conducted if air superiority could be maintained at the front, and an effective defence could be mounted for London. The decision to order reprisals depended on interrelated factors, not least of which was the element of last resort; there seemed to be no way to defend hospital ships or to defend against night attacks by bombers. The other considerations included an appreciation of the military effect (would reprisals actually deter further attacks?), the British capability (would the British air forces be able to mount such attacks and continue to meet their other commitments?), the effect on the morale of the British public and the moral issue of attacking non-combatants.

The government found itself in the unenviable position of facing strenuous opposition whatever it decided. It alone had to deal with the reality of all of the factors – those opposed to reprisal attacks on non-combatants voiced highly emotive moral propositions such as '[d]oes the government think that, if we send aeroplanes to kill little innocent German babies, that is going to help the situation?'[25] without concern for military realities. However, the majority of MPs and press coverage reflected more populist demands for retaliation, without worrying about ethical hazards. This dual pressure does appear to have affected government decisions, albeit in a rather haphazard manner. The usual answer to Parliamentary questions was to plead security and to refuse to discuss the issue,[26] and on the occasions when public engagement was inevitable, the government approach seems to have been apologetic; approving the Freiburg attack, Cabinet decided it 'should be followed by a carefully prepared statement of justification by the Admiralty, explaining that no further reprisals would be taken as soon as the Germans ceased their attacks'.[27]

In presenting the strategic objectives of aerial bombardment, the government was perpetually defensive, vacillating between apologist justification of offensive operations after the event and defensive justification of failures in air defences. The agenda of any public debate was always set by those criticising the government action.

A Cabinet report in August 1917 concluded '[a]nd the day may not be far off when aerial operations with their devastation of enemy lands and destruction of industrial and populous centres on a vast scale may become the principal operations of war'.[28] Almost immediately, the Air Staff started building the concept of operations against populous centres, arguing that, for effect, such a campaign required to be concentrated through large numbers of aircraft attacking near-simultaneously.[29] This work led to the creation of the Independent Force whose focus was to be the aerial bombardment of German towns. In command was General Trenchard who, with much of the 'traditionalist' High Command, continued to argue vociferously against this objective, saying for instance '[t]he main object of the aerial forces should be action in battle[;] their action against the interior of Germany, although of undoubted importance from the point of view of economics and morale, can only be secondary'.[30]

Although the decision to bombard German towns was taken on 2 October, Winston Churchill at the Ministry of Supply was not convinced of the utility of bombing civilian populations: 'It is improbable that any terrorization of the civil population which could be achieved by air attack would compel the government of a great nation to surrender ... In our own case we have seen the combative spirit of the people roused, not quelled, by the German air raids.'[31] This was at odds with the notes of the Cabinet meeting on 2 October: 'it was pointed out that the public, and in particular the poorest classes ... were tending to give way to panic'.[32] Cabinet decided that the Prime Minister should meet principal newspaper editors to 'restrain them from publishing detailed descriptive accounts and pictures ...'[33] Similarly, 'the Prime Minister impressed on General Trenchard the importance [success would] have on the moral[e] of the people at home'.[34] In the event, I have found no evidence that the air bombardment on either side was significant in the outcome of the First World War.

The interwar years

The doctrine of strategic aerial bombardment was championed by the Italian General Giulio Douhet, whose 1921 book *The command of the air* challenged the accepted morality and rules of conventional war: 'the battlefield will be limited only by the boundaries of the nations at war and all of their citizens will become combatants, since all of them will be exposed to the aerial offensives of the enemy. There will be no distinction any longer between soldiers and civilians.'[35]

More pertinent for Britain, Douhet's general views were similar to those now held by Lord Trenchard. On assuming command of the newly formed Royal Air Force (RAF) in April 1918, Trenchard had adopted the concept of targeting populous industrial centres as the *raison d'être* of the new force and a central plank in his argument for its continued independence from the other two armed forces.[36] Douhet's 'big idea' was that the use of long-range bombers to attack industrial and population centres would strike a nation's will to fight so decisively that it could win a war independently of the successes of the other military forces: '[f]irst would come explosions, then fires, then deadly gases floating on the surface and preventing any approach to the stricken area ... A complete breakdown of the social structure cannot but take place in a country subjected to this kind of merciless pounding from the air.'[37]

The Hague Conference of 1923 attempted to draw up rules of air warfare: '[a]erial bombardment is legitimate only when directed at a military objective ... The bombardment of cities, towns, villages, dwellings, or buildings not in the immediate neighbourhood of the operations of land forces is prohibited.'[38] None of the attending powers ratified the Conference's draft.

However, the Hague Convention of 1907 (which antedated aerial warfare) had been ratified; it stated '[t]he attack or bombardment, by whatever means, of towns, villages, dwellings or buildings which are undefended is prohibited'.[39] Debating a fatal leak of phosgene gas (already proscribed) from a German factory, Lord Halsbury bemoaned the nature of potential gas attacks on civilian populations, and the inability to prevent them: '[t]he War proved that a determined attack could always get through'.[40] He considered reprisal by aerial bombing: '[i]t is poor consolation that the only answer we can find to the destruction of half civilisation is that we should be able to destroy the other half ...'[41] and continued '[w]e are not in a position for reprisals ... We have not got the machines [aircraft]'.[42]

Interservice views

The debate within the British government and the RAF about the purpose of air forces was continuous and Trenchard was an adamant advocate:

> the object of all three Services is the same, to defeat the enemy nation, not merely its army, navy or air force ... Air power can dispense with that intermediate step, can pass over the enemy navies and armies, and penetrate the air defences and attack direct the centres of production, transportation and communication from which the enemy war effort is maintained.[43]

This could be interpreted as a means of avoiding the hideous attrition associated with trench warfare experienced by all of the Chiefs of Staff (COS) during the First World War. A more cynical interpretation might be that it was an attempt to preserve the independence of the RAF by articulation of a unique role.

At this 1928 meeting of the COS, Trenchard considered arguments for and against aerial bombardment of military targets within cities.

> Among military objectives must be included the factories in which war material is made, the depots in which it is stored, the railway termini and docks at which it is loaded or troops entrain or embark, and in the general means of communication and transportation of military personnel and material ... What is illegitimate, as being contrary to the dictates of humanity, is the indiscriminate bombing of a city for the sole purpose of terrorising the civilian population ... Moral effect is created by the bombing in such circumstances but it is the inevitable result of a lawful operation of war – the bombing of a military objective.[44]

Trenchard's view was not shared by the other two services. The Chief of the Naval Staff (Admiral Madden) wrote in response 'it is taken for granted that direct air attack on the centres of production, transportation and communication must succeed in paralysing the life and effort of the community and therefore of winning the war. No evidence has so far been produced that such bombing in the face of counter attack will have such a result.'[45] The Chief of the Imperial General Staff (Field Marshal Milne) was similarly critical:

> whilst the Memorandum sets out to show that all the objectives suggested are legitimate military ones, the result in practice would be that, though the objective might be a given boot factory, the actual target would be the town in which the factory happened to be located, and the victims would be its unarmed inhabitants.[46]

Both Madden and Milne criticised Trenchard's proposals because there was no empirical evidence of the effect of bombing and because it did not appear to comply with existing notions of strategy. This was egregiously service-biased; the result of concentration of effort in an industrial-age war had been the defensive entrenchment of the autumn of 1914 and the inability to break the deadlock. The

First World War had demonstrated the fundamental impact that even the primitive aircraft of the era had had, and neither the Royal Navy nor the British Army seem to have been willing to explore the inevitable potential. Clearly both Madden and Milne had ethical reservations, with good cause.

Trenchard's evolving doctrine was testing the boundaries of contemporary ethics. The concept of what is now complacently referred to as 'collateral damage' was novel, and there is little to suggest that the doctrine of 'double effect' was generally understood; it was certainly not common currency. In an almost unique but still oblique reference to the concept in 1918, Jourdain quoted the Archbishop of Canterbury, '[who] distinguishe[d] between the wounding and killing of non-combatants incidental to the bombardment of a fortified town as "sometimes inevitable"'.[47] The ethical doctrine of 'double effect' suggests the casualties caused by bombardment may be excused (but not legitimated), but they must not be the purpose of the bombardment. In this context, the principle of proportionality also emerged – the double effect cannot be so overwhelming that the unintended casualties outweigh (or are disproportionate to) the tactical benefit sought.

The bleakest aspect of Trenchard's proposal was the apparently amoral synopsis with which it concluded: 'whatever be the views held as to the legality, or the humanity, or the military wisdom and expediency of such operations, there is not the slightest doubt that in the next war both sides will send their aircraft out without scruple to bomb those objectives which they consider the most suitable'.[48] The 1935 RAF *War Manual* stated '[a]lthough the bombardment of suitable objectives should result in considerable material damage and loss, the most important and far-reaching effect of air bombardment is its moral effect'.[49]

Naturally, this debate was conducted in the utmost secrecy and not in the public domain. However, fiction had already outstripped military capability; in 1908 H. G. Wells had written *The war in the air*, which had seized the public imagination, describing a world war in which aerial bombing would destroy every major city and bring about 'universal social collapse'.[50] As *The War Game* demonstrated (see chapter one), this idea of universal social collapse, or 'breakdown', is a constant refrain in consideration of the effects of nuclear war and will be developed in chapters three and four. Wells followed up this book in 1914 with *The world set free* in which atomic bombs are used to devastate cities during a putative world war set in 1956 (Wells anticipated the destructive potential of atomic energy after Rutherford published his 'Theory of Atomic Disintegration' in 1902; even though Rutherford did not actually split an atom until 1917). In 1926 the novel *1944* portrayed gas attacks devastating London and in 1932 the film *Things To Come* presented the viewer with inexorable waves of enemy aircraft destroying British civilisation.

Each of these portrayals of future war had a pervasive sense of the inevitable vulnerability of the civilian. There was an appreciation that technological advances had moved warfare away from the relatively localised carnage of the battlefield, where only combatants were at risk, to a future in which civilian populations could become the object, or more or less legitimate, unintended casualties, of aerial attack. It was in this mindset that Prime Minister Baldwin argued in Parliament that

> I think it is well for the man in the street to realize that there is no power on earth that can protect him from being bombed. Whatever people may tell him, the bomber will always get through ... The only defence is offence. You have to kill more women and children more quickly than the enemy if you want to save yourselves.[51]

This fatalistic view was archetypical of the generation that had lived through the horrors of the First World War and found itself preparing for a second. Writing in 1944, Spaight (of whom more later) concluded '[i]f there was one subject upon which there was almost universal agreement before the war, it was, first, that another war would be the end of civilisation, and, secondly, that aircraft would be the prime agents in the causation of that end'.[52] If the British public and their leadership had not understood fully the potential for attack by aerial bombardment, the bombing of towns during the 1937 Spanish Civil War and the 1938 Japanese attacks on China seemed to vindicate Baldwin's prediction. *The Times* reported under the headline 'Town destroyed in air attack'.[53] In a very emotive report, *The Guardian* stated 'Town of ten thousand in ruins ... it is not known how many hundreds of people – men, women, and children – have been killed: it may never be known ... Many of the people who raced desperately for the open fields were systematically pursued and machine-gunned from the air by swooping fighters.'[54]

In the Commons, Sir Archibald Sinclair suggested '[Guernica] was not a case in which civilians were killed in the course of ordinary bombardment, but was a deliberate effort to use air power as an instrument of massacre and terrorism'.[55] In the Lords the following day, Viscount Cecil denounced the raid as 'one of the most horrible things that have ever been done'[56] and the Bishop of Winchester concluded 'an appalling outrage against all the laws of civilisation ... horror has been piled upon horror in this war'.[57]

The Times editorial of 28 April suggested that bombardment might not cause the social breakdown that Douhet had suggested:

> Its aim was unquestionably to terrorise the Basque Government into surrender by showing them what Bilbao may soon expect. Yet so far from having that effect, it may even defeat its object. It may merely inspire the proud democrats of Vizcaya with a passionate determination to fight to the end.[58]

While the immediate response from the Basques was resolute, Bilbao did fall to Franco's forces before any bombing raid was launched against it. Cabinet met on 28 April but did not discuss Guernica, although there was a protracted discussion about the naval situation off Bilbao.[59] A week later, the Foreign Secretary reported that the bombing 'had been received with the utmost horror in America, where it was regarded as a practice for the bombing of London and Paris'.[60] Guernica was bombed by fewer than fifty aircraft, all suited to short-range attacks and all carrying small bomb loads.[61]

The Japanese bombing of Canton and Chungking in the summer of 1938 served to exacerbate fears of bombing. Prime Minister Chamberlain reported to the Commons: '[a]s a result of the raids, approximately 450 were killed and 1,000 wounded, and considerable damage was caused to private property ... whatever may have been the objects aimed at, most of the bombs fell on places which cannot be considered as of military importance.'[62] *The Guardian* described the raids in a dispassionately objective report:

Bombs fell a mile away from the government buildings and half a mile from the railway station. There are no factories or anti-aircraft posts within a mile, but several schools, hospitals, hotels, as well as slums in the area.[63]

The impression that London was vulnerable to this type of attack from German forces in 1938 was a serious overestimate of the capabilities of the Luftwaffe, but that mindset seems to have skewed much of British decision-making about the German threat and the British capability. During 1938, the military situation was regularly analysed for the government by the COS who advised:

[German] AA [anti-aircraft] defences and Air Raid precautions are believed to be in a high state of efficiency and readiness, and the large number of German aerodromes confers great flexibility on her air forces which could concentrate at short notice in any area required.

It must be remembered that Germany's air power is at present considerably greater than our own.

In view of the political objections to the initiation of any action by the Allies which might be misrepresented as an attack upon the civil populace, and the fact that the great initial advantage in air striking power which Germany possesses constitutes a potential menace to the security of this country, we consider that on balance we should be unwise to initiate air attacks upon industrial targets in Germany.[64]

One modern commentator concludes '[a]ir raid phobia – a combination of the perceived offensive advantage in the air and the overestimated German air power – was what powerfully deterred British intervention [at Munich]'.[65] It certainly seems to have figured in the post-Munich debate in both Houses of Parliament. 'I think that perhaps the greatest tribute which can be paid to the work of the Prime Minister lies in the fact that this Debate is taking place to-day under the peaceful conditions that now obtain, and not to the accompaniment of the roar of falling bombs, the fear of which was so present in the mind of every one of us,' said one MP.[66]

Perhaps more illuminating is Viscount Sankey:

I was sorry to read in a paper of repute the other day that had it not been for Mr. Chamberlain's intervention German aeroplanes would have been raining bombs over London. Yes, but it might have added that British aeroplanes would have been raining bombs over Cologne and Berlin.[67]

Sankey reflected both the fear of German attack, and the prevalent ignorance of the real state of RAF capability to retaliate. Cabinet, informed by the COS papers, shared the fear that the Germans would be able to inflict significant British casualties, but also believed that the UK had inadequate defences and was not in a position to retaliate in kind at all. This certainly seems to support the conclusion that '[i]n 1939, the British people expected both to be bombed and to bomb others'.[68]

The fourteen RAF contingency air plans for Europe prior to the outbreak of war included:

> Plans for attacking enemy's manufacturing resources in the Ruhr, Rhineland and Saar (WA5) ... Plans for attacking enemy's air manufacturing resources in Germany (WA6) ... Plans for the attack on specially important depots or accumulations of warlike stores other than air, in enemy country (WA8) ... Plans for attack on enemy's headquarter and administrative offices in Berlin and elsewhere (WA13).[69]

This was very much in line with Trenchard's views on offensive operations and complied with the letter of the 1923 Hague Convention. Despite this doctrinal emphasis on bombing strategic targets in Germany, however, the RAF did not have the capability to achieve any of these plans; no long-range aircraft capable of carrying substantial bombloads; no tactical doctrine for the protection of bombers against fighter defences and, crucially, no means of accurate navigation over long distances other than visual; clearly this was even more limiting at night. The implications of the difference between pre-war expectation and the reality is critical to the evolution of British thinking on, and conduct of, strategic bombing.

Notes

1 Carl von Clausewitz, *On war* [Kindle edn] (Princeton, NJ: Princeton University Press, 2008), 483.

2 John Baylis and Kristan Stoddart, *The British nuclear experience: the roles of beliefs, culture and identity* (Oxford: Oxford University Press, 2014), 18.
3 *The Guardian*, 'Zeppelin raid on London' (14 October 1915), p. 7.
4 *The Guardian*, 'Zeppelin commander on his task' (25 September 1915), p. 6.
5 J. Macpherson, HC Deb. 7 June 1917, *Hansard*, vol. 94 cols 353W.
6 Diana Preston, *A higher form of killing: six weeks in World War One that forever changed the nature of warfare* (London and New York: Bloomsbury, 2015) [Kindle edn], location 42215.
7 Cabinet Office. Minutes of a meeting of the Imperial War Cabinet, 3 April 1917. TNA CAB 23/40/07.
8 Cabinet Office. Minutes of a meeting of the Imperial War Cabinet, 12 April 1917. TNA CAB 23/40/09.
9 HC Deb. 19 April 1917, *Hansard*, vol. 92 col. 1815.
10 HC Deb. 25 June 1917, *Hansard*, vol. 95 cols 318–79.
11 HC Deb. 24 April 1917, *Hansard*, vol. 92 cols 2224–5.
12 Bob Peason, 'More than would be reasonably anticipated: No. 3 Wing Royal Naval Air Service' (2013), available: www.overthefront.com/over-the-front-journal/sample-articles/more-than-would-be-reasonably-anticipated [accessed 15 December 2013].
13 *The Guardian*, 'Hospital ship crimes; naval airmen bomb German town' (17 April 1917), p. 5.
14 *The Guardian*, 'Shrapnel for women and children' (28 May 1917), p. 5.
15 B. Law, HC Deb. 14 June 1917, *Hansard*, vol. 94 col. 1136.
16 *Ibid*.
17 Cabinet Office. Minutes of a meeting of the War Cabinet, 14 June 1917. TNA CAB 23–4–10.
18 *Ibid*.
19 *The Guardian*, 'Aeroplane raid on London' (14 June 1917), p. 5.
20 *The Times*, 'The air attack on London' (14 June 1917), p. 7, col. B.
21 *Guardian*, 'Aeroplane raid on London', p. 4.
22 *The Times*, 'Air raid warnings for the City' (18 June 1917), p. 8.
23 Cabinet Office. Letter from Mr A Baker (Hertford) to Sec., War Cabinet with resolution of a m[ee]t[in]g held in Hertford on 26 Jun[e]. TNA CAB 24/17/75. Cabinet Office. Letter from Mr G. J. Allen (Croydon) to Sec., War Cabinet with resolution of a m[ee]t[in]g held in Croydon (GT-1373, dated 26 June 1917). TNA CAB 24/17/74. Cabinet Office. Letter from Dr Macnamara MP; petition on reprisals from constituents of Camberwell, 9 July 17. TNA CAB 24/17/76.

24 Cabinet Office. Minutes of a meeting of the War Cabinet, 7 July 1917. TNA CAB 23/3/26.
25 S. Collins, HC Deb. 18 June 1917, *Hansard*, vol. 94 cols 1419–21, 1419, 1420.
26 B. Law, HC Deb. 14 June 1917, *Hansard*, vol. 94 col. 1136.
27 Cabinet Office. Minutes of a meeting of the Imperial War Cabinet, 12 April 1917. TNA CAB 23/40/09.
28 Cabinet Office. Second report of the Prime Minister's Committee on Air Organisation and Home Defence Against Air Raids. 17 August 1917 (the Smuts Report). TNA AIR 1/515/ 16/3/83.
29 Air Staff original paper on objectives, 3 September 1917. TNA AIR 1/462/15/312/121.
30 Hugh Trenchard, repr. in H. A. Jones, *The war in the air*, Appendices: 'Being the story of the part played in the Great War by the Royal Air Force' (London: Naval & Military Press 2002), App. X.
31 *Ibid.*, App. IV, 18–22.
32 Cabinet Office. Minutes of a meeting of the War Cabinet, AM 2 October 1917. TNA CAB 23/4/17.
33 *Ibid.*
34 Cabinet Office. Minutes of a meeting of the War Cabinet, PM 2 October 1917. TNA CAB 23/4/18.
35 Giulio Douhet, *The command of the air* (Washington, DC: Air Force History and Museum Programme, 1921; 1998), 10.
36 Tami Davis Biddle, *Rhetoric and reality in air warfare: the evolution of British and American ideas about strategic bombing 1914–1945* (Princeton, NJ: Princeton University Press, 2002), 62.
37 Douhet, *Command of the air*, 58.
38 Hague Conference, 'Rules concerning the control of wireless telegraphy in time of war and air warfare' (1923), available: www.icrc.org/applic/ihl/ihl.nsf/Article.xsp?action=openDocument&documentId=3876F3A2997A8103C12563CD00518519 [accessed 18 December 2013]. Art. XXIV.
39 Hague Conference, 'Convention (II) with respect to the laws and customs of war on land and its annex: regulations concerning the laws and customs of war on land' (1907), available: www.icrc.org/applic/ihl/ihl.nsf/Treaty.xsp?action=openDocument&documentId=CD0F6C83F96FB459C12563CD002D66A1 [accessed 18 December 2013]. Art. 25.
40 HL Deb. 11 July 1928, *Hansard*, vol. 71 cols 963–86, 975.
41 *Ibid.*, col. 974.
42 *Ibid.*, col. 975.

43 Hugh Trenchard, 'Memorandum by the Chief of the Air Staff for the Chiefs of Staff Sub-Committee on the war object of an air force, 2nd May 1928', repr. in Charles Webster and Noble Frankland, *The strategic air offensive against Germany 1939–1945*, Vol. IV, Appendices (London: HMSO, 1961), 71–6.
44 *Ibid.*
45 C. Madden. 'Note by the Chief of Naval Staff for the Chiefs of Staff Sub-Committee on the memorandum of the Chief of the Air Staff, 21st May 1928', repr. in Webster and Frankland, *Strategic air offensive IV*, 81–3.
46 G. Milne, 'Note by the Chief of the General Staff for the Chiefs of Staff Sub-Committee on the memorandum of the Chief of the Air Staff, 21st May 1928', repr. in Webster and Frankland, *Strategic air offensive IV*, 79–81.
47 Margaret Jourdain, 'Air raid reprisals and starvation by blockade', *International Journal of Ethics*, 28 (1918), 542–53: 547.
48 Trenchard, 'Memorandum by the Chief of the Air Staff' repr. in Webster and Frankland, *Strategic air offensive I*, 71–6.
49 Richard Overy, *The bombing war: Europe 1939–1945* (London, Penguin, 2013), 48.
50 H. G. Wells, *The war in the air: and particularly how Mr. Bert Smallways fared while it lasted* (London: George Bell and Sons, 1908), 312.
51 S. Baldwin, HC Deb. 10 November 1932, *Hansard*, vol. 270 col. 632.
52 James Molony Spaight, *Bombing vindicated* (London: Geoffrey Bles, 1944), 2.
53 *The Times*, 'The tragedy of Guernica: town destroyed in air attack' (28 April 1937), p. 17 col. G.
54 *The Guardian*, 'Town of ten thousand in ruins' (28 April 1937), p. 11.
55 A. Sinclair, HC Deb. 28 April 1937, *Hansard*, vol. 323 col. 318.
56 HL Deb. 29 April 1937, *Hansard*, vol. 105 cols 84–92.
57 *Ibid.*
58 *Times*, 'Tragedy of Guernica', p. 17 col. G.
59 Cabinet Office. Conclusions of a meeting of the Cabinet, 28 April 1937. TNA CAB 23/88/07.
60 Cabinet Office. Conclusions of a meeting of the Cabinet on 5 May 1937. TNA CAB 23/88/08.
61 Paul Preston, *The destruction of Guernica* (London: Harper Press, 2012), [Kindle edn], location 189.
62 N. Chamberlain, HC Deb. 3 June 1938, *Hansard*, vol. 336 cols 2411–3.

63 *The Guardian*, '1,100 now dead in Canton' (31 May 1938), p. 11.
64 Committee of Imperial Defence. Appreciation of the situation in the event of war against Germany (COS 764), 13 September 1938. TNA CAB 53/41 (pp. 81–95), paras 11, 23, 28.
65 Gerald Lee, 'I see dead people: air-raid phobia and Britain's behaviour in the Munich crisis', *Security Studies* 13 (2010), 42: 265.
66 A. Southby, HC Deb. 3 October 1938, *Hansard*, vol. 339 cols 40–162, 114.
67 Sankey, HL Deb. 3 October 1938, *Hansard*, vol. 110 cols 1297–366, 1347.
68 M. Connelly, 'The British people, the press and the strategic air campaign against Germany 1939–45', *Contemporary British History* 16 (2002), 2: 44.
69 Committee of Imperial Defence. War plans: report by the Joint Planning Sub-Committee (COS 781), 25 October 1938. TNA CAB 53/41 (pp. 268–85), 16–17.

3

Government, public and total war (1940–45)

At the start of the Second World War, there was little appetite for British forces to be involved in bombing operations against civil targets. Before his Munich conference with Hitler in June 1938, Chamberlain told the Commons:

> In the first place, it is against international law to bomb civilians as such and to make deliberate attacks upon civilian populations ... In the second place, targets which are aimed at from the air must be legitimate military objectives and must be capable of identification. In the third place, reasonable care must be taken in attacking these military objectives so that by carelessness a civilian population in the neighbourhood is not bombed.[1]

This drew directly from the articles of the 1923 Air Power Convention, which had been unanimously confirmed by the League of Nations in 1938. On 1 September 1939 President Roosevelt issued a public appeal to each of the hostile governments:

> publicly to affirm its determination that its armed forces shall in no event and under no circumstances undertake bombardment from the air of civilian populations or unfortified cities, upon the understanding that the same rules of warfare will be scrupulously observed by all their opponents.[2]

The British government responded publicly:

> it was already the settled policy of His Majesty's Government, should they become involved in hostilities, to refrain from such action and confine bombardment to strictly military objectives on the understanding that those same rules will be scrupulously observed by all their opponents.[3]

Given the fear of the perceived German offensive dominance, this position suited the British strategically as well.

Lord Mottistone sought a perpetual commitment to this position shortly after the war had started: 'do [the government] adhere absolutely to this document?'[4] arguing that '[t]his mutual destruction can have no effect on the result of the war' and constructing his case against bombing cities on this assumption. In reply, Lord Halifax (Foreign Secretary) only indirectly challenged the assumption that bombing cities was of no value:

> [with] the fusion of all the activities of a nation, military, civil and industrial, into one gigantic war effort, the distinction to be drawn between combatant and non-combatant becomes vastly more difficult to draw in these days, and to maintain.[5] [Halifax continued] the restrictions that His Majesty's Government have imposed upon the operations of their own forces were based upon the condition of similar restraint being observed by their opponents, and His Majesty's Government must hold themselves completely free, if such restraint is not in fact observed, to take such action as they may deem appropriate.[6]

The next day, the press picked up on this open right of retaliation, while throwing in a swipe at the immorality of likely German actions:

> Until the proposed atrocities are actually committed, the British Government, with the French, will certainly not depart from the assurance that both have given to the President of the United States that they will 'confine bombardment to strictly military objectives on the understanding that those same rules will be scrupulously observed by all their opponents'. But Lord Halifax, in replying to Lord Mottistone yesterday, gave a clear warning that the promise is only binding under the condition stated. If the enemy do not in fact observe the same restraint, then we must retain our own complete freedom of action.[7]

This debate, of course, echoes almost exactly the 1917 reprisal debate. Lord Strabolgi was concerned about the descent from a high moral intent as war progressed:

> we are only in the early days of this war. Passions have not, on the whole, been aroused, and those of us who remember what happened

in the last war will be well aware that, before very much longer, natural but bitter feelings will arise in people's minds and there will be demands for reprisals and demands for giving what is called 'a taste of their own medicine' to our enemies and so on.[8]

He was quite right; the War Cabinet decided on 10 May 1940 that the government would 'now publicly proclaim that it reserves to itself the right to take any action which it considers appropriate in the event of bombing by the enemy of civil populations'.[9] The proclamation was reported verbatim in *The Times* the following day. The fight for the moral high ground over the bombing of non-combatants had already started.

By the start of the Second World War, the pervasive view held by the public and in government was that bombers would always get through and were capable of wreaking such damage on a society that they could cause its breakdown. There was very clear aversion to this being a tenable means for Britain to wage war, in government, in press coverage and among the public at large, but it appears that Strabolgi was right in his assessment that as the war progressed, its realities would overcome these moral restraints.

During the Battle of Britain, Churchill (ever the pragmatist) perceived that the only way to take the offensive against Germany was 'an absolutely devastating, exterminating attack by very heavy bombers from this country upon the Nazi homeland. We must be able to overwhelm him by this means, without which I do not see a way through.'[10] In September 1940, the focus of German air attack shifted from RAF airfields to towns and cities. The Defence Committee of the War Cabinet discussed retaliatory bombing of German towns: Air Chief Marshal Sir Cyril Newall (Chief of the Air Staff) 'strongly favoured a continuation of our policy of attacking military objectives with bombs. Many of these objectives would be near centres of population, and by attacking them we should affect the morale and living conditions of the German public.'[11]

Churchill subsequently wrote:

> About the same time the enemy began to drop by parachute numbers of naval mines of a weight and explosive power never carried by aircraft before. Many formidable explosions took place. To this there was no defence except reprisal. The abandonment by the Germans

of all pretence of confining the air war to military objectives had also raised this question of retaliation. I was for it, but I encountered many conscientious scruples.[12]

This suggests, in a way that official records might not, that there were extensive discussions involving 'scruples'.

Newall's intervention suggests a very particular interpretation of Chamberlain's 1939 assurance that British forces would never be ordered to bomb civilian targets; the civilian casualties and effect on morale and living conditions would not be the purpose of the attacks, merely a side-effect of attacks on military targets. This was a simple but carefully constrained extrapolation of the 'double effect' doctrine advocated by Trenchard in 1928.

Although a member, Attlee was not present at this session of the Defence Committee and later wrote of the decision to bomb cities: '[o]f course the ultimate responsibility for the bombing policy lay with the Cabinet, and I don't seek to evade it, but I thought that concentration on strategic targets such as oil installations would have paid better ... German morale stood up to it pretty well.'[13] British morale generally stood up to the 'Blitz' pretty well too. Churchill records meeting survivors of one raid during a visit in September 1940: '[w]hen we got back into the car a harsher mood swept over this haggard crowd. "Give it 'em back," they cried, and "Let them have it too." I undertook forthwith to see that their wishes were carried out; and this promise was certainly kept.'[14]

Throughout the war, bombing policy was dictated by the Defence Committee. There is a historiographical debate about the influence on, and culpability for, the bombing policy of subordinate commanders, particularly Air Marshal Harris, Commander-in-Chief Bomber Command 1942–5. This debate appears to be founded on an overestimate of the influence on policy, and independence of command, of 'field commanders'. Harris's own frustration at his prescribed scope for initiative is very clear from the correspondence with the Air Staff in the winter of 1944 and suggests that bombing policy was effectively dictated by Cabinet. A February 1942 Ministry of Economic Warfare memorandum describes how policy was made:

> The general aims of bombing are laid down by the Defence Committee with the advice of the Chiefs of Staff. These aims are embodied in a directive which is sent to Bomber Command together with the recommendations of the Air Staff regarding the particular objectives which should be attacked in order to achieve the aims.[15]

Three years later, Harris was characteristically blunt about this level of specific direction and the lack of independence of his command:

> On policy and strategy, I take my instructions at present from the D/CAS[Deputy Chief of the Air Staff] ... In fact I now hold only the tactical, technical and administrative command of a force where operations are otherwise dictated, virtually ad hoc, by the climate, the Air Ministry, SHAEF [Supreme Headquarters Allied Expeditionary Force] and enemy reactions – in that order of impact and import.[16]

Despite the apparently well-defined pre-war doctrine and the fourteen bombing plans, initial efforts were not well coordinated. The primary bombing targets were oil, the aircraft industry, the aluminium industry and railway communications. In Cabinet, Prime Minister Anthony Eden and Defence Minister A. V. Alexander had argued for the bombing of the German people, as some form of retribution, but in January 1941 the priority was shifted to oil, with large industrial cities and communications as alternatives.[17] Two months later, the priority shifted again to submarine building yards and ancillary industries, factories assembling long-range aircraft and naval bases in Germany and occupied territory.[18]

In July 1941, the Air Ministry produced 'Bomber Command', a 130-page pamphlet which described Bomber Command's offensive against the Axis to date. Clearly it was propaganda; British crews were capable of considerable physical and mental endurance; brave yet cautious; cool, yet daring:

> Bomber Command was to leap across the protective barrier of his armies and strike at his vital centres, so as to destroy his factories and oil refineries and to disrupt his communications – in a word, to dislocate and bring ruin to his military economy.[19]
>
> Many a tribute was paid to the accuracy of our bombing. It was said on all sides ... that the British only attacked military objectives and that anyone not living in their neighbourhood was in no danger.[20]

The pamphlet also addressed morale: '[w]hat then has been the effect of our raids on the morale of the Germans? The importance of this aspect of our bombing attacks on them needs no emphasis ...'[21] This statement appears to have made explicit the assumption that bombing legitimately targeted the morale of the population.

John Spaight had been a senior civil servant in the Air Department, retiring in 1937 as Principal Assistant Secretary. He had published regularly on air warfare, becoming an acknowledged authority, and collaborated with Basil Liddell-Hart in the production of the 1938 pamphlet series 'The next war', writing 'The next air war'.[22] Although never writing publicly for the Air Ministry in an official capacity, Spaight remained remarkably well informed and he published prolifically throughout the Second World War. In a 1940 journal article he wrote:

> Night-flying raiders groping for a particular factory or military establishment will probably have to plaster the whole area with bombs ... The indiscriminate bombing which such a method of trying to hit a given objective would involve is hardly a practice which either side will be eager to initiate. Not only would the effect upon neutral opinion be unfavourable to the belligerent who began it, but the result would inevitably be to stiffen the determination of the nation which figured as the first victim of such an attack.[23]

At the time, 'groping' was entirely consistent with RAF capability. The pre-war Cabinet had been concerned about the vulnerability of British civilians to indiscriminate German bombing. The intent for British strategic bombing accorded with Trenchard's view that the bomber would 'attack direct the centres of production, transportation and communication from which the enemy war effort is maintained'.[24] There was a clear expectation that precision bombing would enable this policy while minimising civilian casualties, and this was the public policy. Debilitating casualties in daylight operations forced the RAF to switch to night operations, but these were fraught with navigation problems so the RAF could not attack specific targets with sufficient accuracy to avoid significant civilian casualties and realise Trenchard's ambition.

Instructions to Bomber Command in July 1941 reflected this reality:

you will direct the main effort of the bomber force, until further instructions, towards dislocating the German transportation system and to destroying the morale of the civil population as a whole and of the industrial workers in particular.[25] [Detailed orders included] successful attack of a specific target at night can only be undertaken in clear moonlight. It follows therefore, that for approximately ¾ of each month it is only possible to obtain satisfactory results by heavy, concentrated and continuous bombing of large working class and industrial areas in carefully selected towns.[26]

This view was supported in Cabinet throughout the summer of 1941: 'Mr Butler said that the Secretary of State for Foreign Affairs felt strongly that no departure should be made from our present policy of attacking civilian morale in Western Germany'.[27] That said, however, the War Cabinet also decided not to retaliate against Italian air raids on Alexandria harbour, concluding 'that these were not directed against the civilian population but had been aimed at the harbour which was a legitimate target'.[28]

A report to the Air Staff in August 1941 detailed the lack of bombing precision; only between 10 per cent and 40 per cent of bomber aircraft actually reached the designated target, and of those, only 30 per cent dropped their bombs within five miles of the target.[29] After that report, the Joint Intelligence Committee added 'primary targets selected for night bombing were marshalling yards, centres of rail and water transport, submarine bases and shipbuilding yards [and after February 1942] night bombing [was] devoted primarily to selected industrial areas, attacks on particular factories being attempted only when conditions are perfect'.[30]

A memorandum by the Secretary of State for Air in February 1942 recommended:

> a greater bombing effort be made against Germany for the following reasons (i) This is the time of year to get the best results from concentrated incendiary attack ... (iii) The coincidence of attacks with Russian successes would further depress German morale ... (iii) A new navigation aid is about to come into service ... I therefore recommend (a) that the heavy bomber force be employed without restriction until further notice on the attack of industrial areas and selected precise targets in North-West Germany.[31]

It stipulated Essen, Cologne, Duisburg, Dusseldorf and Gelsenkirchen as the primary industrial areas to be targeted. Accordingly, the bombing directive of 14 February 1942 instructed: '[i]t has been decided that the primary objective of your operations should now be focussed on the morale of the enemy civil population and in particular of the industrial workers'.[32] Air Chief Marshal Portal (Chief of the Air Staff) wrote the next day to his staff: 'I suppose it is clear that the aiming points are to be the built-up areas, not, for instance, the dockyards or aircraft factories where these are mentioned in Appendix A. This must be made quite clear if not already understood.'[33] This is a clear direction to Bomber Command to attack areas, not the specific military targets contained within them.

Spaight's public commentary on the 'war in the air' was updated in 1941:

> the incursions of the Royal Air Force into Germany and of the Luftwaffe into Britain have steadily increased in frequency and vigour. Those of the British airmen have been aimed exclusively at impairing Germany's military strength. Oil refineries, synthetic oil plants and petrol storage depots have been among the chief targets ... [RAF] pilots and bomb-aimers had been trained to a pitch not even approached by those of the Luftwaffe. Precision of aim was inculcated and practised. Long periods were spent in the search for and exact location of targets. If the designated objective could not be found, and if no alternative target could be bombed with reasonable precision, no attack was launched.[34]

However, whilst this was consistent with the declared policy, it was no longer an accurate reflection of the operations themselves.

Under the headline 'Bombs on Berlin', the *Daily Mirror* had reported on the RAF raid on Berlin in September 1940: '[t]ypical of RAF bombing operations, the raid was made by a force of bombers which delivered their attack with great precision'.[35] The editorial went on to state that an offensive policy was the only one acceptable: '[t]he air war is no time for lecturers and gloved persons wishing to live up to a high standard of ancient chivalry. The invention of the bombing plane abolished chivalry for ever. It is now "retaliate or go under"'.[36] The broadsheets were no less committed to supporting the efforts of Bomber Command. In *The Guardian*'s view, '[i]n war, reprisals can be exacted only for want of a policy ... The purpose of bombing, as of all military action, is to bring the war to a successful end as quickly as possible.'[37] This, of course, is a complete derogation from *The Guardian*'s 1915 condemnation of 'haphazard murder'.[38] In only eighteen months the press had achieved the descent from moral high ground to amoral passion that had so concerned Lord Strabolgi.

There is little evidence of public dissent at this stage, and what there was, was disregarded. When the pacifist MP Richard Stokes challenged the bombing policy, the Government response was '[t]hat is exactly what Hitler would like'.[39] And when asked in November 1941 about the 'Committee for the Abolition of Night Bombing' the Home Secretary responded 'I have no reason to suppose that this misguided propaganda is attracting or will attract any serious attention'.[40] This suggests that the dominant view was that to question the morality of the bombing campaign would in turn harm the British war effort. However, the Archbishop of Canterbury, opening the Upper House of the Convocation of Canterbury on 27 May 1941, stated:

> It is one thing to bomb military objectives, to cripple the industries on which the prosecution of the war depends, and, alas! in so doing it may be impossible to avoid inflicting loss and suffering on many civilians. It is a very different thing to adopt the infliction of this loss as a deliberate policy ... I do not believe that the great majority of British folk, even in the bombed areas, really want such a policy. It is to be hoped that the Government will resist any pressure and make it clear that they will adhere to their declared policy.[41]

The Church's moral endorsement was obviously absolutely conditional on the assumption that the RAF was engaged in precision bombing and that civilian casualties were the unavoidable by-product.

In May 1942 Mr Justice Singleton was commissioned to conduct an independent investigation into the effect of the bombing policy:

> In the light of our experience of the German bombing of this country, and of such information as is available of the results of our bombing of Germany ... If an industrial area, which has as its centre important factories engaged on war work, is taken and dealt with thoroughly by concentration of bombing better results are likely to be achieved.[42]

Singleton recognised that technical limitations had a serious impact:

> Unless and until a greater measure of accuracy in target finding can be reached it will probably prove to be good policy to keep chiefly to targets which can be found fairly easily. [And he summarised] the effect on Germany's war production and effort will be very heavy over a period of twelve or eighteen months, and such as to have real effect on the war position.[43]

Area bombing?

Singleton's report provided independent verification of an earlier Joint Intelligence Committee report which had also considered the salient military factors affecting the bombing mission: damage to factories, the need to hold back a number of fighter aircraft to attack Allied bombers, keeping occupied a large number of men and guns on anti-aircraft work and on searchlights and a very large number on air raid precautions, and the lowering of morale.[44] The COS were able to draw on monthly estimates of the effects of bombing on Germany's war effort, and the bombing policy was regularly altered accordingly. Lord Cherwell, Churchill's chief scientific advisor, advocated area bombing:

> Investigation seems to show that having one's house demolished is most damaging to morale. People seem to mind it more than having their friends or even relatives killed. At Hull signs of strain were evident, though only one-tenth of the houses were demolished. On the above figures, we should be able to do ten times as much harm to each of the fifty-eight principal German towns. There seems little doubt that this would break the spirit of the people.[45]

His position was supported fully by the Secretary for the Air, Sinclair: '... we see no reason to doubt that within eighteen months, and with American help, the degree of destruction which Lord Cherwell suggests is possible can, in fact, be achieved'.[46]

Having considered the salient tactical factors in his assessment of the effect of the bombing, Singleton concluded '[t]he question of morale is a much more difficult one with which to deal. There has been no break of morale in this country, although some people think that there was a danger of it locally on one or two occasions when bombing suddenly ceased.'[47] British Air Intelligence had surveyed the effect of the German bombing campaign on the UK in August 1941 and reported that 'no town in England has suffered a breakdown in morale'.[48]

With RAF aircraft now capable of carrying over five tons of explosives into Germany, Singleton continued:

> Now we expect to be able to deliver to Germany in the future a much greater weight of bombs than we received ... I doubt whether our bombing ability is, or in the near future can be, sufficient to bring about a break in morale in this way alone. Herein again arises the importance of increased accuracy. I prefer to think of the effect on morale combined with the other factors and envisage the bombing of an industrial area with important factories in the centre, rather than the bombing of houses, and I think better results will be achieved thereby.[49]

Of course the industrial areas to which he referred tended to be situated in densely populated districts. This sophisticated position, very precisely between the deliberately indiscriminate bombing of cities for the purpose of 'dehousing' and the precision bombing of 'military-industrial' targets, was probably the most accurate depiction of British bombing policy after 1942. It was exquisitely poised on the knife edge of the 'double effect' and collateral damage issue discussed above.

A very small minority of MPs queried the policy of area bombing as the British campaign developed. In 1941 Stokes had dismissed night bombing as 'contagious lunacy'.[50] In March 1943, he pressed home his questions: 'my constant objection has been to what I call indiscriminate night bombing – and bombing must be indiscriminate at night'.[51] He challenged Under-Secretary of State for Air Balfour to reconcile his assertions about not bombing indiscriminately with Sinclair's more recent statement that 160 acres of built-up accommodation areas around Essen had been destroyed in attempts to hit Krupp's works. Balfour replied;

> Of course, war is cruel and destructive, and the destruction of property and cities is inevitable, but again, I give the assurance that there is no change in our policy, that our purpose is to destroy Germany's

industry, transport and war industry and war potential, and that we are not wantonly bombing women and children for the sake of doing so.⁵²

Stokes pushed his point again three weeks later, asking Sinclair 'whether on any occasion instructions have been given to British airmen to engage in area bombing rather than limit their attention to purely military targets?' to which Sinclair replied '[t]he targets of Bomber Command are always military, but night bombing of military objectives necessarily involves bombing the area in which they are situated'.⁵³

In his memoirs, Harris wrote that he had been perturbed at the difference between public and operational policy: 'I personally thought this was asking for trouble; there was nothing to be ashamed of, except in the sense that everybody might be ashamed of the sort of thing that has to be done in every war, as of war itself'.⁵⁴ The Air Staff had responded to a question he had raised that:

> No attempt has been made to conceal from the public the immense devastation that is being brought to the German industrial cities ... the widespread devastation is not an end in itself but the inevitable accompaniment of an all-out attack on the enemy's means and capacity to wage war ... It is, in any event, desirable to present the bomber offensive in such a light as to provoke the minimum of public controversy and so far as possible to avoid conflict with religious and humanitarian opinion. Any public protest, whether reasonable or unreasonable, against the bomber offensive could not but hamper the Government in the execution of this policy and might affect the morale of the aircrews themselves.⁵⁵

This of course is a very particular interpretation of Portal's blunt instructions, and clearly irritated Harris who justifiably felt that there was no place in the conduct of his missions for the moral sophistry being practised in Whitehall. Target priority varied and Bomber Command provided support for the D Day landings and attacked Ministry of Economic Warfare priorities such as oil refineries or ball bearing production, but it was Singleton's position that Harris argued against with Portal in late 1944: '[a]rea bombing must enter into any scheme, because in bad weather we have to use sky markers, we must have a large target ... and we necessarily in those conditions paint with a large brush'.[56]

Harris subsequently wrote that 'the policy of destroying industrial cities, and the factories in them, was not merely the only possible one for Bomber Command at that time, it was also the best way of destroying Germany's capacity to produce war material. The morale of the enemy under bombing could be taken as an imponderable factor.'[57] He was clearly committed to the policy of area bombing of industrial targets; in this, he was inclined to follow Trenchard's 1928 doctrine. He does not appear to have been philosophically committed to morale as a target; in the efforts to destroy industrial capacity, the workforce would become casualties.

Opposition to the RAF strategic bombing campaign

In 1944, the small, but vociferous, opposition to the bombing campaign published 'The seed of chaos'. This pamphlet described the campaign as 'obliteration bombing' and quotes a 'foreign observer' writing for the *News Chronicle*:

> the principle behind your raids is to make sure of hitting important targets by wiping out the whole area in which they lie ... This is obviously the only way to get results in a highly industrialised country like Germany, but it also has the advantage of producing an automatic effect on the population ... There is a definite point at which the weight of bombs dropped in a certain area in a certain time produces no longer a fitful feeling of alarm ... but an unbearable strain, more or less approaching panic.[58]

This is very much the position that Portal had already taken with Bottomley, and was exactly in accord with Harris's understanding of his directives, but it was not a position that the British government openly acknowledged. Shortly afterwards, Spaight published a remarkable book analysing the bombing campaign: *Bombing vindicated*. In it, he considered this domestic opposition:

> German propagandists were able to count upon a certain amount of support in their campaign from within this country of free speech. That it was the support of only a tiny fraction of the population was

shown when, on 29 April 1942, Mr Rhys Davies, a Pacifist Member of Parliament, questioned the Secretary of State for Air about the recent raid on Lübeck and implied in a supplementary question that the air offensive should be stopped. There was a resounding cheer throughout the House of Commons when Sir Archibald Sinclair replied: 'The best way to prevent this destruction is to win the war as quickly as possible'. A few weeks later, another Socialist Member, Mr R. R. Stokes, was asking about the recent 'thousand-bomber' raids and their utility.[59]

Spaight dismisses Davies and Stokes as a pacifist and socialist respectively and not representative of the population. But what is most interesting about his book is that it exists at all. The book is an extremely well-informed and entirely sympathetic portrayal of the public bombing policy, associated with a successful campaign 'narrative'. In effect, it addresses the opposition argument, in terms in which ministers might not want to become embroiled; it is the government case by proxy. Spaight observes that:

> What has really happened is that air power has killed absenteeism in war ... We are all in the thick of the trouble now. Naturally, to those who have not grown accustomed to being no longer absentees it is nothing but an intrusion, a trespass, a violation, an outrage, when war thus invades their hearths and homes ... This, they cry, is not war – it is murder. But it is war – the new kind of war. It is wrong, horrible, unendurable, but it was inevitable. It was inevitable that the air offensive against an enemy's source of armed strength should come and with it the incidental killing of non-combatants.[60]

This is what Harris had wanted the Air Staff to admit and Sinclair to announce, but he was not politically astute enough to recognise that the official position could never reflect this bleakly amoral position.

When the political direction is perceived as somehow covert, or dishonest a commander of armed forces personnel faces a problem: how can those personnel be assured of the moral justification of their mission if it is misrepresented and not acknowledged in public because it might have difficult political repercussions? I experienced this discomfort in command during the 2003 invasion of Iraq; I was (and remain) entirely content I could justify the mission to myself and to my ship's company, but not in terms of which I felt the government would approve. Harris was in the same position.

The lessons learned from the strategic bombing campaign

In the event, predictions of the crushing moral effect of aerial bombardment were as overstated as the 1937 *Times* Guernica editorial had suggested; instead of Douhet's social breakdown, the populations of the cities that did endure persistent attacks drew together and perversely gathered strength. The German attacks of 1940–41 never achieved the levels of destruction that the Royal Air Force would inflict two years later. Even the most destructive operations of the European war, the protracted attacks on Hamburg, Berlin and Dresden, did not realise Douhet's social breakdown.[61] That said, immediate post-war military analysis suggested that the effect on German morale of these devastating attacks had been significant; Erhard Milch, the Nazi State Secretary for Air, had told the members of his ministry '[i]f we get just five or six more attacks like these on Hamburg, the German people will just lay down their tools, however great their willpower'.[62]

The actual effect of the bombing campaign, as opposed to that perceived in the UK in 1946, is not pertinent here, but Milch's observation was reported in time to inform the next iteration of British strategic planning. Those making decisions about Britain's atomic weapon capabilities in the decade after the end of the 1942–5 bombing campaigns had all held senior positions during the Second World War. The lessons that they derived from that conflict would inevitably colour their thinking as they shaped atomic policy.

The British authorities started the Second World War with an exaggerated fear of a German strategic bombing campaign and sought to avoid provoking it. That said, this does not appear to be an early example of a coherent deterrence strategy. Although Attlee later considered that in the 1930s 'strategic bombing had been hailed as the great deterrent to war and had failed',[63] Trenchard, Baldwin and Halsbury were not considering deterrence by posing a capability to respond in kind; simply that attacks on cities were inevitable in future conflict and, in order to defeat an enemy, Britain must be able to inflict and survive such damage.

Similarly, deterrence was not discussed in the Commons debate on Munich in September 1938 – the assumption was that war would inevitably entail bombing of cities. The RAF strategic bombing campaign was simply the manifestation of that thinking. The capability offered by strategic bombing, realised much more by the RAF than by the Luftwaffe, pushed the boundaries of the laws and ethics of war: as a 1940 *Daily Mirror* headline suggested, 'Maybe everybody is a victim'.[64]

The RAF bombing campaign was limited by the inadequate preparations made between the wars. Having started with an exaggerated expectation of precision bombing of military targets, Cabinet accepted less discriminate bombing of areas in which military targets were situated and then specifically directed operations against the morale of the German civilian population. The issue of unintended civilian casualties caused by attacks on military targets forced consideration of the legitimacy of targets located within residential areas. The directives to Bomber Command were explicit – the primary objective of operations was to be the morale of the enemy civilian population. Despite diversions into oil and other 'panacea' targets, this was essentially the policy until the end of the war. Ministers consistently denied this in Parliament and in public.

It is clear from the public interventions of religious leaders and the paucity of philosophical debate on the topic beforehand that ethical issues of proportionality and non-combatant immunity were novel, highly contentious and only marginally developed. An embryonic, but highly articulate and energetic opposition emerged during the war – not simple pacifism but a much more sophisticated opposition to specific aspects of modern warfare. The small group of men who comprised the highest level of government, the Cabinet Defence Committee, had to address these issues, in the context of a war of unprecedented potential which posed an existential threat to the UK. They chose to avoid the inevitably contentious and morale-sapping public debate about the difference between 'unintended but inevitable casualties' and deliberate targeting of non-combatants.

They clearly believed they had no alternative if they intended to continue to prosecute the war; they lied.

Notes

1. N. Chamberlain, HC Deb. 21 June 1938, *Hansard*, vol. 337 cols 937–8.
2. *The Times*, 'Air bombing of civilians' (2 September 1939), p. 10.
3. *Ibid.*
4. Mottistone. HL Deb. 13 September 1939, *Hansard*, vol. 114 col. 1047.
5. Halifax. HL Deb. 13 September 1939, *Hansard*, vol. 114 cols 1050–52, 1050.
6. *Ibid.*, 1052.
7. *The Times*, 'Bombing of open towns' (14 September 1939), p. 9.
8. Strabolgi, HL Deb. 13 September 1939, *Hansard*, vol. 114 col. 1052.
9. Cabinet Office. Minutes of a meeting of the War Cabinet, Friday 10 May 1940. TNA CAB 69/1: 4.
10. Churchill, *Their finest hour*, [Kindle edn] location 9798.
11. Cabinet Office. Defence Committee (Operations), Minutes of a meeting of the War Cabinet, Tuesday 24 September 1940. TNA CAB 69/1: 204.
12. Churchill, *Their finest hour*, location 5419.
13. Francis Williams, *Twilight of empire: memoirs of Prime Minister Clement Attlee* (New York: A. S. Barnes & Co., 1962), 49.
14. Churchill, *Their finest hour*, locations 9798–804.
15. Ministry of Economic Warfare, 'Night bombing as an instrument of economic warfare' (4 February 1942) briefing by O. L. Lawrence, repr. in Webster and Frankland, *Strategic air offensive IV*, 214–19.
16. Air Staff, Letter, Harris to Portal ATH/DO/4 (J) (18 January 1945). DEAN 02/10 (ATH/DO/4) Liddell Hart Archives.
17. Cabinet Office. Defence Committee (Operations), Minutes of a meeting of the War Cabinet, 13 January 1941. TNA CAB 69/2: 36–40.
18. Joint Intelligence Committee, 'Effect of bombing policy: report by the Joint Intelligence Sub-Committee' JIC(42)117(D) 6 April 1942. TNA CAB 79/20/16: 128–56, para. 2.
19. Air Staff, 'Bomber Command, the Air Ministry account of Bomber Command's offensive against the Axis' (London: HMSO; September 1939–July 1941), 104.
20. *Ibid.*, 123.
21. *Ibid.*, 122.
22. James Molony Spaight, *Air power in the next war* (London: Geoffrey Bles, 1938).
23. James Molony Spaight, 'The war in the air; first phase'. *Foreign Affairs*, 18 (1940).
24. Webster and Frankland, *Strategic air offensive IV*, 71–6.

25 Ibid., 135–9.
26 Ibid.
27 Cabinet Office. War Cabinet; Chiefs of Staff Committee, Minutes of a meeting held on 14 July 1941. TNA CAB 79/12; 438–40.
28 Cabinet Office. Defence Committee (Operations), Minutes of a meeting of the War Cabinet, 3 July 1941. TNA CAB 69/2; 271–7.
29 Air Staff, 'Report by Mr. Butt to Bomber Command on his examination of night photographs' (18 August 1941), repr. in Webster and Frankland, *Strategic air offensive I*, 205–13.
30 Joint Intelligence Committee, 'Effect of bombing policy', 128–56, paras 3–7.
31 Cabinet Office. Defence Committee (Operations), 'Memorandum on bombing policy' DO(42)14, 9 February 1942. TNA CAB 69-4 pp. 182–3.
32 Air Staff, 'Bombing directive' (14 February 1942), repr. in Webster and Frankland, *Strategic air offensive IV*, 143–5.
33 Reproduced in Martin Middlebrook and Chris Everitt, *The Bomber Command war diaries: an operational reference book, 1939–45* (Harmondsworth: Viking Penguin, 1985): 240.
34 James Molony Spaight, 'The war in the air; second phase'. *Foreign Affairs*, 19 (1941), 408–13: 408.
35 *Daily Mirror*, 'Bombs on Berlin' (12 September 1940), p. 3.
36 Ibid., p. 5.
37 *The Guardian*, 'Heavy bombing' (19 April 1941), p. 19.
38 *Guardian*, 'Zeppelin commander on his task', p. 6.
39 HC Deb. 24 July 1941, *Hansard*, vol. 373 cols 1051–2.
40 HC Deb. 27 November 1941, *Hansard*, vol. 376 cols 886–7.
41 *The Guardian*, 'Primate condemns "retaliation"' (28 May 1941), p. 28.
42 Cabinet Office. Report by Mr Justice Singleton: 'The bombing of Germany' May 1942. TNA PREM 3/11/24 (112): 3.
43 Ibid., 4.
44 Joint Intelligence Committee, 'Effect of bombing policy': 5 para. 4.
45 Cabinet Office. Defence Committee (Operations), 'Memorandum: estimation of bombing effect' DO(42)38, 9 April 1942. TNA CAB 69/2 p. 267 (the De-Housing paper).
46 Cabinet Office. Defence Committee (Operations), 'Memorandum: estimation of bombing effect' DO(42)39, 13 April 1942. TNA CAB 69/4.
47 Cabinet Office. Report by Mr Justice Singleton: 4.
48 Air Intelligence Department. *Air defence of Great Britain*, Vol. III: App. A: Morale. TNA AIR 41/17: 1.
49 Cabinet Office. Report by Mr Justice Singleton: 4.

50 HC Deb. 24 July 1941, *Hansard*, vol. 373 cols 1051–2.
51 HC Deb. 11 March 1943, *Hansard*, vol. 387 cols 922–62, 951.
52 *Ibid.*, col. 954.
53 HC Deb. 31 March 1943, *Hansard*, vol. 388 col. 155.
54 Arthur Harris, *The bomber offensive* (Barnsley: Collins, 1947), 58.
55 Air Staff, Letter, Street to Harris CS.21079/43 (15 December 1943). TNA AIR 14/843.
56 Air Staff, Letter, Harris to Portal ATH/DO/4 (A) (1 November 1944). DEAN 02/10 (ATH/DO/4). Liddell Hart Archives.
57 Harris, *Bomber offensive*, 88.
58 Vera Brittain, 'Seed of chaos: what mass bombing really means' (London: New Vision Publishing Co., 1944), 22.
59 Spaight, *Bombing vindicated*, 34.
60 *Ibid.*, 117.
61 There is a substantial body of work looking at the conduct and ethics of strategic bombing campaign and associated RAF command and control authorities to which I cannot do justice here. I would recommend Biddle, *Rhetoric and reality in air warfare*; Peter Gray, 'The gloves will have to come off: a reappraisal of the legitimacy of the RAF bomber offensive against Germany', *Air Power Review* 13 (2010), 9–40; A. C. Grayling, *Among the dead cities* (London: Bloomsbury, 2006); Victor Gregg, *Dresden: a survivor's story* (London: Bloomsbury Reader, 2013); Keith Lowe, *Inferno: the devastation of Hamburg 1943* (London: Penguin Group, 2007); and Hugo Slim, *Killing civilians: method, madness and morality in war* (London: Hurst, 2007).
62 Air Division BAFO 1946. 'Report on German flak towers, Flak Disarmament Branch'. TNA AIR 55/ 158. Reported in Lowe, *Inferno* [Kindle edn], location 3695.
63 Baylis and Stoddart, *The British nuclear experience*, 18.
64 *Daily Mirror* (12 September 1940).

4

From the Second World War to continuous at-sea deterrence

If the early leaders of Britain's nuclear enterprise learned anything from the experience of the Second World War, it was that 'total' war could threaten the whole fabric of the state and society. The previous chapter suggests that the importance of maintaining a plausible position of moral authority was considered critical in keeping public morale engaged in the war effort, even if the enemy engaged in immoral attacks on the British population. This moral authority was directly linked to the strategies to be used, and therefore to the technical capabilities available; common ethical considerations would limit the extent of destruction acceptable to the public, even if such limits were detrimental to the war effort.

The wartime government had skirted around these issues. In 1945, Attlee's government was faced with the significant challenge of developing weapons specifically designed to destroy entire cities, and strategies to use them, without alienating the public. How this was achieved is the subject of this chapter.

The genesis of the British atomic energy project

In April 1937, opposition leader Clement Attlee requested the Foreign Secretary (Lord Halifax) to take immediate steps to address a protest to Franco and Hitler over the Guernica bombing.[1] Eight years later, as the Prime Minister of the newly elected Labour Government, Attlee was faced with leading the UK into the atomic age when the first atomic bomb was dropped on Hiroshima on 6 August 1945 and the second on Nagasaki three days later. Japan unconditionally surrendered on 12 August 1945. For the victorious Allies, there was no question that the atomic bombs had caused the capitulation.[2]

Attlee intuitively understood the central premise of deterrence. In August 1945, he wrote to the Cabinet '[t]he answer to an atomic bomb on London is an atomic bomb on another great city ... Even the modern conception of war to which in my lifetime we have become accustomed is now completely out of date.'[3] A month later he wrote to President Truman in much the same vein: 'I have so far heard no suggestion of any possible means of defence. The only deterrent is the possibility of the victim of such an attack being able to retort on the victor.'[4]

Attlee's insight was not initially shared by the COS; in response to a Cabinet request to consider whether the 'introduction of atomic explosives open[s] up an era of destruction on a scale never before considered feasible, or is ... merely an intensive development of the existing concept of war'[5] the Admiralty response to the COS concluded that the atomic bomb would not on its own be decisive and that it was merely a bigger and better bomb.[6] This rather missed the point that Attlee seems to have intuitively grasped; one 'bigger and better bomb' carried by a single aircraft could inflict as much damage and as many casualties as the massed raids of the latter part of the Second World War. The atomic bomb embodied the vulnerability of cities and populations that had so agitated Attlee after Guernica; one bomber could always get through. Atomic weapons seemed to provide an affordable means for a perpetual capability to threaten cities that otherwise would have required a war economy fully mobilised for production of huge numbers of heavy bombers.

Initially, Attlee considered international control of atomic energy essential to the maintenance of peace. He 'declared the intention of His Majesty's Government to devote all their efforts to making the new discovery serve the purpose of world peace and to co-operate with others to that end.'[7] In a personal memo for the Cabinet, he enlarged on this: '[a]ll nations must give up their dreams of realising some historic expansion at the expense of their neighbours'.[8] The UK and USA conducted simultaneous negotiations throughout the autumn of 1945, looking both at international control, culminating in the Washington Declaration of 15 November 1945, and at cooperation in the construction of national stockpiles of atomic weapons, culminating in an Anglo-American Memorandum of Understanding the next day.[9]

In extreme secrecy, Attlee created a small Cabinet Committee (GEN 75). As with each subsequent government, all key nuclear policy decisions have been made in a cabinet sub-committee with very restricted membership, which maintained security and also tended to obscure from subordinates the factors considered in reaching these decisions. GEN 75 core membership was limited to the Prime Minister, Foreign Secretary, Chancellor of the Exchequer, Secretary of State for Dominion Affairs, President of the Board of Trade and Minister of Supply. The Minister for Defence was not a core member. At its first meeting, GEN 75 commissioned a report from the newly appointed Advisory Committee on Atomic Energy (ACAE), which initially reported that:

> [t]he political difficulties in the way of securing the adoption of any of the [international control] schemes and of assuring their efficacy are very great and might at first sight seem insurmountable. But the consequences of making no agreement at all are so grave not only for the security of all countries but for our present civilisation itself, that it appears imperative to devise a system of control.[10]

GEN 75 met to discuss international control on average twice a week throughout October 1945 and the final ACAE report of 29 October 1945 concluded that international control was impracticable and that '[n]o international agreement, therefore, is likely to be successful which attempts to restrict the freedom of any of the major powers to produce atomic weapons'. The report went on to recommend that '[t]he United Kingdom Government should itself undertake the production of atomic bombs as a means of self-defence as soon as possible'.[11] A week later GEN 75 agreed that

> [i]t was useless to suppose that if a war once began, any agreement not to use the atomic bomb would be observed. The sanction proposed would never be effective in deterring a country which was in danger of being destroyed in a war begun without the use of atomic weapons ... we must, therefore, base our foreign and defence policy on the assumption that if another war took place, these weapons would be used.[12]

Accordingly, GEN 75 initiated the British atomic energy programme. This conflict between what was clearly a genuine aspiration for a system of international control, and the 'hard power' realism of developing a national atomic capability, is present throughout these early Cabinet discussions and it has been an underlying tension in the formulation of British nuclear deterrence policy ever since.

This decision was not specifically to develop a British atomic weapon, however, merely to develop the facilities to produce the materials required; but the bomb was an implicit objective. Margaret Gowing described the decision to embark on a British bomb project as 'axiomatic'.[13] These men simply assumed that the UK was a top-rank world power with significant strategic and commercial interests. The concurrent (non-atomic) meetings of the Defence Committee in 1946 addressed British strategic issues worldwide from Greece and Denmark to Palestine and India;[14] it was unthinkable that Britain should be without atomic weapons. The other common theme of those Defence Committee meetings was one of severe austerity and reductions in defence manpower, equipment and spending, with Bevin arguing that 'the proposed cuts in the armed forces faced him with grave difficulties in obtaining the support which he thought necessary as a backing to the conduct of the government's foreign policy'.[15]

Once the decision to proceed with a British atomic energy project had been made, the focus shifted to its management; Cabinet discussed national control of research and production, ministerial responsibility, the creation of the appropriate advisory bodies and the executive control of the research, even the site of the proposed research establishment and its organisation and financial structure. Privately, Attlee subsequently explained that he viewed a British bomb as 'essential. We had to hold up our position *vis-à-vis* the Americans. We couldn't allow ourselves to be wholly in their hands and their position wasn't awfully clear always ... The manufacture of a British bomb was therefore at that stage essential to our defence.'[16]

In January 1946, the team to lead the development of a UK weapon was appointed – Air Chief Marshal Portal was to be the

'Controller of Production Atomic Energy' in overall charge of developing the project; Christopher Hinton (an ICI specialist in munitions production) in charge of the construction of the 'pile'; and John Cockcroft (scientist and engineer) the Director of the Atomic Energy Research Establishment. GEN 163, an even smaller group of ministers than GEN 75, met only once (in January 1947) to consider Portal's initial report. It recommended that 'a decision is required about the development of atomic weapons in this country. The Service Departments are beginning to move in the matter and certain sections of the Press are showing interest in it.'[17] Portal's paper went on to offer three options: do nothing, proceed within legacy procurement arrangements, or create a specific structure for the design and development of atomic weapons. The paper's conclusions were couched so as to encourage the latter, and especially the ability to maintain secrecy. GEN 163 formally agreed the development of a British atomic bomb, but this decision merely reflected the growing consensus among ministers and officials.[18]

Importantly, GEN 163 also agreed the structure of the organisation to build the bomb. Security was to be draconian and Hinton's organisation, responsible for producing the radioactive core of the weapon, was not even made aware of the major bomb project (High Explosive Research) set up under William Penney (a physicist recently returned from the USA). Penney's presence in the Armaments Research Division prompted some press speculation, with some wondering whether his appointment heralded the start of a British atomic weapon development programme.[19] Secrecy was maintained within the respective official arms of the project through the Official Secrets Act and Portal's compartmentalisation of the various activities, and media interest was inhibited through the use of the D (Discretion)-notice system.

The D-notice system has no statutory basis and was developed early in the 20th century in order to prevent the press from publishing information which might be of value to a future enemy; it persists today. D-notice 25 inhibited publication of any articles or broadcasts on British atomic weapon activities from 1945 until its replacement in 1971 by a standing D-notice, which in turn was replaced in 2000 by Defence Advisory (DA) Notice 02:

> It is requested that disclosure or publication of highly classified information about nuclear and non-nuclear defence equipment or equipment ... should not be made without first seeking advice ... Release of highly classified operational plans and security arrangements could potentially jeopardise the safety and security of our nuclear forces and reduce their deterrent value.[20]

Famously reticent at the best of times, Attlee actively minimised discussion of a British atomic bomb, but that was equally true of all aspects of defence; in the Lords' discussion of the 1948 Defence White Paper, Viscount Bridgeman lamented: '[o]ur only information is still derived from the *New Statesman and Nation*'.[21] There was no public engagement, nor was the wider Cabinet informed.

In a manner that was to become typical of the release of British atomic project information, Cabinet and the Commons were informed of the existence of a British project later in 1948, not because they were democratically elected representatives with a right to know and a duty to oversee, but because there were concerns that there might be a leak and they would find out by some other route. In the Commons debate on the Defence White Paper in March 1948, one MP pointed out the inconsistency in making the strategic bomber force the major striking force, calling it 'a most wasteful force in terms of economic and industrial work, and ... also a most expensive and an inefficient force for doing damage to the enemy'.[22] However, he had failed to consider what such a force might be carrying, such as atomic bombs. Another stated that 'the time has come, or is coming, when the Prime Minister, or perhaps the Foreign Secretary, should tell us exactly how we stand ... I wonder how many of us ... asked ourselves how we should defend these islands and the Commonwealth in the event of atomic war?'[23] Even during these debates, the evidence that the UK was developing an atomic bomb capability was mounting, but was not connected with the concept of a British deterrent.

In the event, Defence Minister Alexander announced in a response to a question two months later: 'all types of modern weapons, including atomic weapons, are being developed'.[24] Press coverage of the announcement was muted, hardly surprising since the D-Notice Committee had spent much of the previous meeting agreeing the terms of the new D-notice 25 that would cover

'Location and progress of work UK on development and production of atomic weapons, their design, size and other details, materials used, storage locations, identification of individuals with work on atomic weapons'.[25] The government sent D-notice 25 to newspaper editors the day before Alexander's statement. In the absence of an official atomic narrative, the British public and media had simply been using their imagination, fuelled by both fact (mostly from American atomic literature) and fiction.

The public atomic narrative

Early in 1946, Father Siemes's eyewitness account of the aftermath of the attack on Hiroshima was published in the *Irish Monthly*. It was a miserably objective account of destruction, mass selfishness and, occasionally, individual altruism of survivors. It recounted some lucky stories but graphically described a society broken down to the extent that civilised behaviour was almost eradicated within days, and a near-immediate reversion to a state of nature followed: '[t]housands of wounded who died later could doubtless have been saved if they had received proper treatment and care, but rescue work in a catastrophe of this magnitude had not been envisaged; since the whole city had been knocked out at a blow'.[26]

It was followed in August 1946 with the publication of John Hersey's book *Hiroshima* in a dedicated issue of the *New Yorker*: '[t]he American response was sensational and the text was republished in full by several newspapers, ABC radio broadcast a reading of the entire text over four nights, and the book version of the text became an immediate best seller'.[27] It was less of a sensation in the UK; *The Guardian* carried a description of the *New Yorker* story on 3 September, concluding that '[o]ne hopes that before long all over here, and in the world generally, will have the opportunity to read these pages. Man has either to abolish war or to accept Hiroshima's fate for his own city in the future.'[28] The BBC broadcast a reading of the text in four episodes on the Third Programme on the evenings of the week of 14 October 1946. *The Observer*'s critic, W. E. Williams, complained that the programme proved heavy going. In print [Hersey's] narrative manages to conceal some of its weaknesses – its redundancies of testimony for instance and its excessive poverty of language. The reader's eye, adjusting the mind to such deficiencies, can select what is to be skimmed and what is to be pondered. But a reading aloud, even by six voices, denies us this selectivity ... I thought the wireless version of *Hiroshima* inordinately long-drawn-out.[29]

A month later, he complained that '[m]atters of immense topical interest do not necessarily make good radio'.[30] It appears that *The Observer*, although it did provide a critique on contemporary coverage of atomic issues, was more interested in style than substance. *Hiroshima* was repeated in one broadcast on the BBC Light Programme on 30 November. Penguin published the British edition of Hersey's book in November 1946, but to considerably less popular reaction than in the USA.

The dichotomy between the bombing of Hiroshima and the results of the Nuremberg trials, which concluded in August 1946, was not lost on some of the public who wondered how:

> the interests of peace or justice ... could ever be served by the massacre, in circumstances of unspeakable horror, of tens of thousands of defenceless women and children? [We are] continually being told that it is no defence of the soldiers and sailors at Nuremberg that they were merely obeying orders; in fact that they should have disobeyed orders which were clearly opposed to humanity and the laws of war. What then about the airmen who were ordered to drop this token of progress on the nurseries and maternity homes of Hiroshima?[31]

These were the same sort of strongly emotive, binary, moral terms in which in Parliament in 1939 Lords Mottistone and Strabolgi and, in 1917 Mr Molteno, had opposed strategic bombing or reprisals against civil populations, and were similarly difficult to refute without a complex ethical debate on deontological and consequentialist precepts. After the experiences of the strategic bombing campaign of the Second World War, this was not a debate into which Attlee's government was prepared to enter in 1946.

Early days of the British nuclear enterprise

At the end of the Second World War, Britain's nuclear expertise had resided almost exclusively inside the Manhattan Project in the USA. With the 1946 McMahon Act, the USA inhibited any prospect of sharing information or technology with others, including any further exchange with those who had contributed to the wartime atomic bomb project.[32] There was a significant shortage of specialist scientists and engineers; the British atomic projects were competing against each other in the same pool with all other elements of defence research, and some who might have been available were put off by the obsessive security restrictions. This secrecy exacerbated further the bureaucratic tensions inherent in the structure that had been established by Portal. This straddled both the MOD (from whom the atomic bomb project was completely hidden) and the Ministry of Supply, for whom Portal worked (technically, although he reported directly to the Prime Minister).[33] Penney struggled even to recruit draughtsmen and secretaries.[34] To say that the development of the British atomic bomb was achieved on a shoestring budget would be to assign a rather sanguine interpretation to the resilience of the project. The casting of the plutonium core of the bomb be tested at Monte Bello in 1952 was conducted by the project director – who had never cast anything – because his specialist had caught the wrong train and they could not wait for him to arrive.[35]

When the time came to test the British atomic bomb, Churchill (once again prime minister) was typically adamant about the need for close control of all information released: 'he would personally approve any communiques or statements which it was desired to issue to the Press. Apart from these, nothing whatever should be said to the Press and nobody should be given any discretion at all to make any statement of any kind. In matters of this kind the less said the better.'[36]

A government statement was issued in February 1952, only once the convoy of vessels involved in the test had departed the UK. The statement simply said that the UK would test an atomic bomb at some point that year, and that the test would be in Australia. The vacuum in public information prompted a flurry of speculation, particularly in the foreign press which was unencumbered by the D-notice system.

A bomb, however, is only one element of a system; it requires delivery platforms, and the personnel to man them. While the decision was being made to pursue a British atomic weapon, the only readily available delivery platforms were heavy bomber aircraft. Air Staff Operational Requirement (OR) 229, for a replacement for the Lancaster and Lincoln bombers, was placed by the RAF in early 1947, with a due delivery date of 1957: '[a] medium range bomber landplane capable of carrying one 10,000lb (4,500kg) bomb to a target 1,500 nautical miles (1,700mi; 2,800km) from a base which may be anywhere in the world'.[37] Although the capability to deliver an atomic bomb was not specified in the design, it was implicit.[38] As above, the usually perspicacious Commons did not seem to have made the connection between the small production run of OR229 and atomic weapons.

The austerity that afflicted the other branches of the armed forces at this time limited development and production of these aircraft, and it was 1957 before the first fully operational combination of the first UK atomic bomb (Blue Danube) and Valiant aircraft of 138 Squadron arrived at RAF Wittering, still in extreme secrecy. By that stage, however, technology had advanced; the launch of Sputnik in October 1957 very publicly made the point that missiles rendered the bomber at best obsolescent as a single strategic deterrence delivery platform. It was this remorseless technological advance, in particular the advent of the hydrogen bomb and its seemingly limitless destructive potential, that energised a coherent opposition.

The political decision-making which led to the British hydrogen bomb project continued to be conducted in absolute secrecy, with Churchill establishing his own reduced Cabinet Committee (GEN 465) to discuss the factors and options. The Committee agreed that:

> we should need to get clear the fundamental issues of foreign policy and strategy which were raised by the latest developments ... [including] in the light of the new information about the hydrogen bomb, the following points:–

(i) The likelihood of war.
(ii) The form which war was most likely to take if it came.[39]

The full Cabinet agreed on 22 March 1954 that

> [a] general statement on the strategic importance of the hydrogen bomb would certainly provoke a demand for a full debate. It was desirable that, at the appropriate moment, the public should be made aware of the full implications of the development of the hydrogen bomb. But the timing of this disclosure should be carefully judged.[40]

In March 1954 the USA publicly announced the detonation of the first weaponised hydrogen bomb, the Castle Bravo test.[41] The announcement, accompanied with graphic photographs of the first hydrogen bomb test (which had been conducted in 1952), caused an upsurge of interest in nuclear weapons.

The *Daily Mirror* carried a front-page photograph of the explosion under the headline 'The monster'. On the centre pages the headline was 'The horror bomb' and the leader read:

> The ball of death, poised on a stem of fire, has thrust through the clouds to 40,000ft – half as high again as Everest, the world's highest mountain ... Everything within three miles of the explosion was completely wiped out ... 'Light' damage was recorded up to ten miles ... a bomb detonated 20ft above a city would create a much greater area of total damage.[42]

The initiative for proactive engagement over a public announcement had been lost. Richard Crossman (Labour MP and diarist) noted on 6 April 1954, '[a]ctually of course the American bomb has been manufactured since 1950 and the announcement that the Russians had detonated theirs occurred on August 8th last year. But it's only during these last ten days that the country has become H-bomb conscious.'[43] Macmillan noted that '[i]t is obvious that there is tremendous interest, almost panic, in many parts of the world, about the Hydrogen Bomb'.[44]

Under the headline 'Churchill confesses' the *Daily Mirror* pilloried the Prime Minister's prior UK knowledge of the hydrogen bomb tests and reported that '... last night about a hundred [Labour MPs] put their names to a motion demanding government steps to OUTLAW THE BOMB and to BAN FURTHER EXPLOSIONS' [emphasis in original] and asserted that there were 'murmurs of protest from the Labour side when Sir Winston replied that "if such a request would lead to these results, we ought to be careful about asking the question".'[45]

The Labour Party called an adjournment debate in Parliament, ostensibly to support an immediate summit with the UK, USA and USSR to discuss means to control hydrogen bomb development, and Attlee made a generous and bi-partisan speech which was well received by the House. But Churchill resorted to a rather churlish attack on Labour's record. Macmillan noted in his diary: '[t]he press has been bad – though not quite as bad as I feared. The *D Telegraph* loyal; the *Times* insufferably pontifical; the *Express* and *Mail* in full support; the *Chronicle* and *M Guardian* fair; the *Daily Mirror* vile. Yet with all the criticism, I feel that the main strategic purpose has been secured.'[46]

In May, the COS reported to the Defence Committee that 'the world situation has been completely altered by recent progress in the development of nuclear weapons ... A provisional estimate of the effect of 10 bombs dropped one each on 10 selected cities in the United Kingdom indicates ... the death roll would be [between] 5 millions [and] ... 12 millions'.[47]

It was July before Churchill actually informed the full Cabinet that Britain had already embarked on the development of the hydrogen bomb; the reaction of the Cabinet was to walk out:

> [Churchill] told us that the decision had been made to make the hydrogen bomb in England and the preliminaries were in hand. Harry Crookshank at once made a most vigorous protest at such a momentous decision being communicated to the Cabinet in so cavalier a way, and started to walk out of the room. We all did the same and the Cabinet broke up – if not in disorder – in a somewhat ragged fashion.[48]

Cabinet reconvened the next day and the minutes record that '[t]he Cabinet resumed their discussion of the question whether our atomic weapons programme should be adjusted as to allow for the production of thermo-nuclear bombs in this country'.[49] Macmillan's diary records: '[w]e began on the Hydrogen Bomb. PM said that only the first preliminaries were decided. It was, we recognised, a hideous decision ... A short but valuable discussion followed.'⁵⁰

The main discussion points raised included:

> Was it morally right that we should manufacture weapons with this vast destructive power? There was no doubt that a decision to make hydrogen bombs would offend the conscience of substantial numbers of people in this country ... The point was again made that there was no difference in kind between atomic and thermo-nuclear weapons; and that, in so far as any moral principle was involved, it had already been breached by the decision of the Labour Government to make the atomic bomb. It was also argued that the moral issue would arise, not so much on the production of those weapons, but on the decision to use them: ... if we were ready to accept the protection offered by United States use of thermo-nuclear weapons, no greater moral wrong was involved in making them ourselves ...
>
> No country could claim to be a leading military Power unless it possessed the most up-to-date weapons; and the fact must be faced that, unless we possessed thermo-nuclear weapons, we should lose our influence and standing in world affairs ...
>
> Doubt was expressed about the feasibility of keeping secret, for any length of time, a decision to manufacture thermo-nuclear weapons in this country. It was therefore suggested that thought should be given to the question of how a decision to manufacture these weapons should be justified to public opinion in this country and abroad.⁵¹

These key questions would inform policy development for the remainder of the 1950s. The defence budget was under perpetual strain and the cost of nuclear weapon development has been a significant factor debated ever since. The moral question was raised formally in Cabinet for the first time, although clearly the ministers and officials had long felt it keenly. Indeed, the Cabinet Secretary's notes of the meeting record: 'PM: Must take a decision in principle; not necessarily today; doesn't depend on technical detail; mainly a moral question'.⁵² Macmillan had noted in his diary that 'Churchill broods a good deal about the atomic and hydrogen bomb. The destructive power of the latter is frightful. All London in one night.'⁵³ This moral reasoning appears to have been deeply personal and recorded only in diaries or letters; it was certainly not addressed formally in committee.

The implications for British influence in the world were clearly articulated, and sustained the belief that nuclear weapons conferred status that would otherwise be denied to the UK. The COS reported '[t]he danger that the United States might succumb to the temptation of precipitating a "forestalling" war cannot be disregarded. In view of the vulnerability of the United Kingdom we must use all our influence to prevent this.'[54] This is an early and clear indication of the genuinely strategic implications of this oft-derided motive for an independent British nuclear deterrent.

Cabinet endorsed the plan to continue development of an independent British hydrogen bomb and the British project was announced to Parliament in March 1955. In the meantime, work had begun to address some of the issues raised by the hydrogen bomb threat. The COS report in May highlighted that

> [a] small number of the latest nuclear weapons can achieve a devastating effect ... means of delivery against which there is no foreseeable defence will be developed. These two factors are creating a new military situation which will reduce progressively the value of certain conventional war preparations and weapons.[55]

Similarly, the Committee on Defence Policy had already reviewed the strategic assumptions underlying the current defence policy and the scale and pattern of military (and civil) defence programmes. It concluded '[o]ur primary aim must be to prevent a major war. To that end we must strengthen our position and influence as a world Power and maintain and consolidate our alliance with the United States.'[56] To that end, the Committee proposed 15 per cent cuts to the army and navy, and changes to the air defence structure of the RAF. It then continued:

> The policy outlined in this report will clearly need most careful presentation to the public. Many people are preoccupied with the destructive power of the latest atomic weapons. Fewer perhaps have yet recognised that the development of these weapons may have made major war less likely. The public as a whole will therefore find it difficult to understand why, as the destructive power of air attack increases, we propose to cut down our fighter and anti-aircraft defences and reduce the scale of our expenditure on Civil Defence. These and other changes recommended in this report certainly could

not be defended in isolation. Public acceptance of them can only be secured if they are presented as parts of a coherent plan based on the recognition that no purely defensive policy could ensure the safety of these islands and those who live in them and that the main weight of our defence effort must now be concentrated on building up the deterrent strength which will prevent the outbreak of a major war.[57]

The report could not have been more explicit on the need for a coherent public engagement plan and it proposed a special Defence White Paper in the autumn of 1954 in order to present a full statement of the new defence policy as a whole.

The Committee on Defence Policy report also highlighted the issues associated with civil defence in the event of a nuclear attack. It pointed out that:

[we] should devise alternative means of maintaining the essential machinery of government with maximum flexibility and devolution ... It will not be realistic to attempt to provide shelter for the civil population ... except for certain priority classes, no considerable section of the population could be moved from the main target areas.[58]

The Joint Intelligence Committee and GEN 465, under the guidance of the Cabinet Secretary, commissioned William Strath, director of the Cabinet Office Central War Plans Secretariat, to conduct a study to consider the effects of a nuclear attack on the UK, a requirement that was endorsed by Macmillan and Churchill in December 1954.[59] The Strath Report painted a bleak picture of the British state after a nuclear attack:

there might be complete chaos for a time and civil control would collapse. In such circumstances the local military commander would have to be prepared to take over from the civil authority responsibility for the maintenance of law and order and for the administration of government ... He would have to direct the operations of various civil agencies, including the police, the civil defence services and the fire service. In areas less badly hit the civil authorities might still be able to retain control but only with the support of the armed services.[60]

The importance of this report for defence planning is difficult to overemphasise. The phenomenon of the failure of civil society was abbreviated to 'breakdown' and became the object of very highly classified study in itself: 'breakdown might be defined as occurring when the government of a country is no longer able to ensure that its orders are carried out. This state of affairs could come about through breakdown of the machinery of control ... or of morale.'[61] This was the fracture in the fabric of society which had been predicted by H. G. Wells in 1908 and had influenced strategic decision-making throughout the Second World War.[62]

The Strath Report informed decision-making in Whitehall but only within circles constrained by the very highest levels of discretion and secrecy. Despite the very clear Committee on Defence Policy recommendation, no public engagement was undertaken. In particular, the Report's conclusions were not distributed to the local authorities, where it was assumed that they would stimulate questions about shelter policy and evacuation which the government was ill-equipped to answer.[63] Defence planning continued to develop nuclear deterrence as the core of British defence policy, culminating in the Sandys defence review of 1957. The Strath Report, which in effect went underground in Whitehall, drove expenditure away from civil defence – what was the point? – and was very influential for decades to come. As described above, the issues Strath had described so graphically were central to Peter Watkins's film *The War Game* a decade later, but remained a truly vexed issue for government. As will be considered later, the Thatcher government was to run afoul of exactly the same issues nearly thirty years later; and to a great extent, they remain pertinent in the 2020s, especially to a society which has experienced first-hand the debilitating effects of a pandemic, the associated panic buying, imposition of severe state inhibitions on personal freedoms and increasingly volatile protests against it.

Anti-nuclear opposition

There had been 'peace' movements in Britain since the end of the First World War. They tended to be small and regional but came together in their opposition to the strategic bombing campaign of 1943–5. Although the post-war Labour Party, while in government, had commissioned the British atomic energy project, it was fundamentally split on the issue of atomic weapons, a division that surfaced as one of a number of issues which became characterised by the Bevanite/Gaitskellite divide after the party's defeat in the general election of 1951.[64]

The Peace Pledge Union established a non-violence commission 'to study and discuss the possibility of direct action to seek withdrawal of American forces, stoppage of the manufacture of atomic weapons in Britain, withdrawal of Britain from the North Atlantic Treaty Organisation (NATO), and disbandment of the British Armed Forces'.[65] Although the Union itself achieved little public traction, a number of these groups started to cooperate and the National Council for the Abolition of Nuclear Weapons Tests managed to draw together a number of regional groups opposed to nuclear testing. Opposition within Parliament against the manufacture and possession of atomic weapons was not really organised until the debate about the British hydrogen bomb in 1954 which led to the Hydrogen Bomb Committee: '[t]his was not specifically unilateralist; it was an attempt to see nuclear weapons as a problem of foreign policy … It later led to CND [the Campaign for Nuclear Disarmament].'[66]

This movement was lent momentum by the Suez crisis and the invasion of Hungary, and it was given specific focus by the prevalence of nuclear weapons in the defence policy announced in the Sandys defence review of 1957. The Labour Party struggled to define its position on atomic weapons. Bevan, a committed campaigner for unilateral disarmament, sponsored a motion at the 1957 Labour Party conference calling for unilateral action to end the British atomic weapons programme. In the event however, in a now infamous *volte-face*, when it came to the conference, he actually said:

> I know that you are deeply convinced that the action you suggest is the most effective way of influencing international affairs. I am deeply convinced that you are wrong. It is therefore not a question of who is in favour of the hydrogen bomb, but a question of what is the most effective way of getting the damn thing destroyed. It is the most difficult of all problems facing mankind. But if you carry this resolution and follow out all its implications and do not run away from it you

will send a Foreign Secretary, whoever he may be, naked into the conference chamber. Able to preach sermons of course; he could make good sermons. But action of that sort is not necessarily the way in which you take the menace of this bomb from the world ...[67]

The substance of Bevan's speech remains a core contention in modern arms control debate, and the motives behind it remain obscure. In the event, 'it was his ghost which would continually return to haunt the relationship between Labour and CND throughout the next thirty years'.[68] The day after Bevan's speech, the USSR launched Sputnik 1, demonstrating missile technology that almost instantly made Britain's bomber-based atomic deterrent obsolescent. In December 1957, Harold Steele, a Quaker backed by the 'Direct Action Committee' (DAC), tried unsuccessfully to disrupt the first British H-bomb test. The DAC established an objective of non-violent civil disobedience, and after successful demonstrations at potential missile bases, their first major project was a march from London to Aldermaston, planned for Easter 1958.

The 1957 Reith Lecture series, titled 'Russia, the atom, and the West',[69] was given by George Kennan, ex-US Ambassador to Moscow. It was an examination of contemporary strategic factors, and as with any complex argument, the points Kennan made were deeply nuanced and interrelated. Selective quotation could, and did, support less scrupulous arguments for either side of the disarmament debate but members of the public were able to listen to the full series, and make up their own minds; potentially fully informed of the issues.

The government simply did not engage with the CND movement. Although set-piece events such as the Aldermaston marches seized the popular imagination at the time, their impact on opinion does not appear to have been long-lived; certainly little enduring political momentum was generated. *The Times*, in an unusual editorial, considered that:

> [t]he nation is engaged on a great debate. It is a debate on fundamental issues ... on the outcome of which, it can for once be said without extravagance, depends our civilisation ... Only if there were a sudden wave of hysteria, or a complete national loss of judgement, could any future enemy or present ally fail to see the essential determination

of Britain. The temper of the great national debate is, therefore, all important. It is not a debate to be carried on with histrionics or dramatics. And let it be said that Mr MACMILLAN, Mr GAITSKELL, Mr BEVAN and other political leaders have brought to it a seriousness and purpose that has lost nothing by remaining quiet.[70]

In March 1958 R. A. Butler, the Home Secretary, appeared on BBC's investigative television programme *Panorama* to discuss Soviet nuclear testing. *The Times* reported: '[h]e exposed himself to a crossfire of questions from five accomplished controversialists who bitterly oppose the Government's basing of defence policies on the big bombs ...'[71] Already, television was portraying this complex issue in a manner designed at least as much to provoke dramatic argument on camera as to elicit public understanding of the issue.

The Executive Committee of what became CND met in January 1958, a month before the first CND meeting in Westminster Hall. Although noted in *The Times* diary page,[72] and very well attended, it was not reported subsequently. After the success of this inaugural meeting, the executive committee revised its initial policy statement:

We shall seek to persuade the British people that Britain must:

(a) renounce unconditionally the use or production of nuclear weapons and refuse to allow their use by others in her defence;
(b) use her utmost endeavour to bring about negotiations at all levels for agreement to end the armaments race and to lead to a general disarmament convention;
(c) invite the cooperation of other nations, particularly non-nuclear powers, in her renunciation of nuclear weapons.[73]

From the start of the movement, there was a tension between the CND executive, which was convinced of the need to influence the political process through the conversion of the Labour Party to a unilateral disarmament position, and many of the 'rank and file' and the DAC, who saw the 'movement as essentially extra-parliamentary'.[74] On 4 April 1958, 4,000 anti-nuclear demonstrators met in Trafalgar Square to march to Aldermaston in a march organised by the DAC and endorsed by the CND. The report in *The Times* the following day was a simple record of fact; there was no hyperbole and it was ostentatiously non-partisan. It concluded: '[t]he politicians, who have a vested interest in mass movements, have been disturbed lately by the signs of an emotional popular approach to the problems of nuclear weapons'.[75] The *Daily Mirror* report focused more on the walkers, observing that none of Canon Collins, Foot, Mikardo or Dr Soper (high-profile CND activists) had actually completed the march but had remained at home and rejoined it in Reading.[76]

In 1959, 20,000 met the march from Aldermaston at Trafalgar Square in London but CND was unable to convert this hard core of activists into a coherent political force. The crisis over the erection of the Berlin wall between June and November 1961 raised international tensions and when both the USA and USSR resumed nuclear testing in the middle of the crisis, they lent the nuclear war issue a real air of urgency. Anti-nuclear demonstrators took full advantage of this and non-violent direct action by the 'Committee of 100', a radical group led by philosopher Bertrand Russell, led to thirty of its leaders being imprisoned for a month. In 1962, five of the leaders were imprisoned for eighteen months for breaches of the Official Secrets Act.[77]

The influence of CND 'peaked' in early 1960. CND's magazine *Freedom* called upon activists to 'make [the 1960 Easter demonstration] the biggest demonstration Britain has ever seen ... In this way we might finally get rid of nuclear weapons.'[78] The Labour Party conference of 1960 adopted a resolution supporting unilateral nuclear disarmament. CND General Secretary Peggy Duff described the vote as 'one of the highlights of our campaign', whilst John Cox (later CND Chair), said '[f]or a while it seemed that CND would soon succeed in changing the country's nuclear policies'.[79] Driven by a few of the largest trades unions, against vehement opposition from the leadership and the constituencies, there was a serious division within the Labour Party with Gaitskell and Callaghan advocating a pro-NATO anti-unilateralism line, and Foot, Crossman and Castle supporting the CND position. The combined block vote of the five largest unions defeated two-thirds of the Parliamentary Labour Party and a majority on the National Executive Committee, despite Gaitskell's speech which culminated in:

> What sort of people do you think we are? Do you think we can simply accept a decision of this kind? Do you think we can simply become overnight the Pacifists, Unilateralists and fellow-travellers

that other people are? ... I ask delegates who are still free to decide how they vote to support what I believe to be a realistic policy on defence, which could yet so easily have united this great Party of ours, and to reject what I regard as the suicidal path of unilateral disarmament which will leave our country defenceless and alone.[80]

After concerted effort on the part of the Labour Party National Executive over the next twelve months, the vote was reversed the following year. Although the matter at hand was the vote for unilateral nuclear disarmament, Crossman was convinced that the crux of the issue was more about whether the '... Labour party can be run by personal or collective leadership. That's the real issue – not defence',[81] and he told Gaitskell as much on the night after the vote. Whether Gaitskell took Crossman at his word or not, ironically, the 1960 defeat in Scarborough had actually strengthened Gaitskell's position as leader of the Labour Party and his overwhelming victory against Wilson in the ensuing leadership contest prevented the more left-wing and pro-disarmament lobby from attaining executive power or dividing the party prior to the 1964 general election.[82]

The second generation of the British strategic nuclear deterrent

Throughout the 1950s, British efforts to establish a national atomic deterrent capability progressed, despite significant cost constraints. The British Nuclear Deterrent Study Group (BNDSG) was established in the MOD in July 1959 to '... consider how the British controlled contribution to the nuclear deterrent can most effectively be maintained in the future, and to make recommendations'.[83] Its draft interim report concluded that the programme to replace free-fall bombs (Blue Danube) with the air-delivered short-range Blue Steel missile would remain credible only for a short period. These were the criteria for assessment: were missiles required due to the vulnerability of manned bombers? Would ballistic missiles continue to be less vulnerable than cruise-type missiles?[84] Given the UK's geographic position, would missiles need a mobile platform? What would be the earliest date by which aircraft or submarine systems could be deployed, allowing for assistance from the USA? Was there a gap between this date and the projected end of the service life of the V-bombers,[85] and; if so, what delivery systems were available that might bridge the gap?[86]

There was then a series of separate meetings considering the merits of different delivery systems. By April 1960, the group's revised interim report considered that 'from the mid-1960s onwards, the only weapons system which would give a reasonable assurance of maintaining a significant deterrent capability in all circumstances would need to consist of ballistic missiles launched from either seaborne or airborne long endurance mobile platforms'.[87] Silo-based weapons in the UK could not be sited far enough apart for survivability against attack without encroaching on built-up areas.[88] After the failure of the British Blue Streak missile project, Britain joined the US 'Skybolt' programme in June 1960.[89] This programme's cancellation in December 1962 left British plans for an independent nuclear deterrent in disarray.

However, this public perception (on both sides of the Atlantic) may have been managed, rather than real.[90] The British government had known since before the original agreement that the Skybolt programme was high-risk. The BNDSG had been scoping the generic technical requirements for a nuclear deterrent system that would remain in service and credible into the 1970s, and invulnerability to surprise attack was an increasingly significant factor in the conceptual discussions.[91] The Royal Navy began exploring the potential for a submarine-based deterrent in 1958.[92] A 1961 report by Flag Officer Submarines suggested the Royal Navy could purchase and operate submarines armed with American Polaris ballistic missiles. Although finding little support amongst the Chiefs of Staff at the time, the report was presented to the BNDSG and Defence ministers.[93]

The US government was seeking an operating base for its own Polaris submarines on the west coast of Scotland, as part of an overall 'deal' for providing US assistance to the British deterrent modernisation. As part of the Skybolt deal, Cabinet agreed in principle to provide operating and maintenance facilities at Holy Loch on the Clyde for US Navy Polaris submarines.[94] Integral to these ongoing negotiations was that '[UK] proposed to the United States government that they should offer us a simple option to buy POLARIS submarines if at any time we wished to do so, in return for the facilities we provided in Scotland'.[95]

Contained with this (very high-level) contingency planning was a consideration of the public messaging that would be required if the US Navy were to station Polaris submarines on the Clyde; the main points of concern were the level of control over Polaris missile launch that the British government could exert, and a comparison of the vicinity of US population centres to nuclear operating and storage facilities. In the event, the USA was unwilling to compromise on the control of launch at sea, and the Cabinet agreed 'it would be preferable that the proposed public statement should not indicate that there could be no certainty of consultation in an emergency'.[96]

Macmillan noted on 16 December 1962, on the eve of his trip to meet President Kennedy at Nassau, that '… I got back to a meeting on Skybolt and Polaris which lasted till late in the evening. We shall have a difficult time with the Americans in Nassau.'[97] Clearly he was expecting to negotiate hard, but the tenor of his thoughts on Nassau suggests that, rather than the public debate about purchase of American air-launched or submarine-launched systems, the real debate was about command and control of those nuclear forces:

> Broadly, I have agreed to make our present bomber force (or part of it) and our Polaris force (when it comes) a NATO force for general purposes. But I have reserved absolutely the right of HMG to use it indefinitely 'for supreme national interest'. These phrases will be argued and counter-argued. But they represent a genuine attempt (wh[ich] Americans eventually accepted) to make a proper contribution to interdependent defence, while retaining the ultimate rights of a sovereign state. This accepts the facts of life as they are. But I do not conceal from myself that the whole concept will be much knocked about by controversy at home.[98]

The arrangements made by Macmillan and Kennedy to enable the British purchase of the US Polaris system were agreed by Cabinet on 3 January 1963 and the Polaris Sales Agreement was signed in April 1963. Macmillan was unsure of the reception his 'deal' would receive:

Whether Parliament and the country will think we have done well or badly I cannot tell yet. Yesterday's press was quite good (except of course Lord Beaverbrook's).[99] Today's (Sunday's) is very bad. The Opposition will attack our whole record on defence ... The 'Patriots' (led by Lord Beaverbrook and the isolationists) will accuse us of 'selling out Britain'. No one will find it profitable to take a fair and balanced view.[100]

In the ensuing debate, he was proved generally right. In the meantime, the world moved on. Far from highlighting the dangers of nuclear weapons, the Berlin Wall crisis and the Cuban missile crisis seemed to suggest that deterrence worked, since the superpowers were seen to back away from the 'brink' and the compelling urgency that had given such an edge to the CND campaign simply ebbed away.

Cuban missile crisis

One area where prominent political leadership and careful government public presentation of British nuclear deterrence policy and strategy might be expected to be uniquely salient would be in the midst of an international crisis with a nuclear dimension. To date, fortunately, there has been only one, and on that occasion political leadership and careful government presentation in the UK were conspicuously lacking.

The first Macmillan knew of the Cuban missile crisis was in the evening of Sunday 21 October 1962, when he received from President Kennedy a short account of a serious situation which was developing over Cuba. This was followed by a visit from the United States' Ambassador David Bruce, who provided by hand a 'long letter from President Kennedy, as well as a great dossier to prove that ... there had now been secretly deployed in Cuba a formidable armoury of MRBMs [medium-range ballistic missile] and IRBMs [intermediate-range ballistic missile] which were a pistol pointed at America ... and which could not be tolerated'.[101] Having spoken with Home (Foreign Secretary) and Maudling (Chancellor), Macmillan spoke to President Kennedy by telephone after the latter's televised presidential address to the nation on Monday

22 October. The full text of the address was published on Tuesday 23rd in most of the British press, with no apparent context or input from the British government.

The Times simply reported the full text of the address; the *Daily Express* reported the speech but also ran front-page coverage of the deployment of US forces into the Caribbean, the evacuation of families from the US base at Guantanamo Bay and the surprise apparently expressed by members of the Cabinet at the lack of notice given to the UK.[102] *The Guardian* also reported the full text under the headline 'Soviet deception on Cuba' and ran a front-page story: 'Missile bases built up – 1500-mile range claimed' and an editorial: 'Caribbean storm warning'.[103]

Parliament was in recess until 25 October, when the Queen's Speech was due to open the next session. Cabinet met on the morning of 23 October and the traditionally very concise (to the point of terse) minutes concluded that 'the country would expect to be informed at once of the government's reaction to President Kennedy's speech ... The Leader of the Opposition, who had been informed of the situation on 21st October, had given no undertaking to refrain from condemning the United States action and some of his supporters would almost inevitably do so.'[104] There was no discussion of an early recall of Parliament nor a proactive public statement. This may be due to an assumption that informing the country was synonymous with informing Parliament, and in any case the government had nothing to say; '[s]umming up the Prime Minister said that it was clear that no decision on policy could be taken until some firmer indication had been obtained of the probable nature of Mr. Khrushchev's reaction to the United States blockade of Cuba'.[105] The Foreign Office did issue a short statement:

> Her Majesty's government feel deep concern at the provocative action of the Soviet Union in placing offensive nuclear weapons in Cuba. Mr Gromyko lately gave the most positive assurance that the weapons which the Soviet Union was supplying to Cuba were purely defensive ... Instructions have been sent to Sir Patrick Dean (permanent British Representative to the United Nations) to support the American request to the Security Council that they should make recommendations to end this danger.[106]

As a statement of fact, the Foreign Office statement lacks little, but it leaves the way ahead to the imagination of the reader, or the editor.

In contrast to the restrained and very factual press reports of Tuesday 23 October, Wednesday's *Daily Mirror* reported under the headline 'US hunts target no. 1 – the red rocket runner ... American warships ... are lying in wait for the Soviet rocket-carrying freighter *Polotavia*'. Also on the front page, it reported that leave had been stopped for all Soviet armed forces personnel, and on the back page it reported 'a massive strike force of American jet fighters is being assembled'.[107] The editorial commentary 'Courage, but is it wise?' considers that Kennedy had acted with great courage over Cuba but then concludes:

> What Kennedy has done is to bring the cold war to a head. Now it is Mr Krushchev's turn to try to frighten us. He has begun by alerting the Red Army. We shall hear warlike words. He may take action in Berlin ... for although Mr Kennedy had every right to declare a blockade ... the repercussions may be world-wide. If they are, they cannot fail to affect us.[108]

On 25 October, Cabinet met once more and decided '... it did not seem that there was any action that the Prime Minister could usefully take at the present juncture; an early visit to Washington, for example, could easily be misinterpreted as a mission of appeasement'.[109] There was no discussion on public messaging. That afternoon, Parliament met and Macmillan made a statement to the House simply outlining the facts and presenting an oddly ambivalent position: purely tacit support for the American position. He advocated support to the UN efforts to resolve the situation, but when challenged by Gaitskell refused to be drawn to overt support for the US actions:

> In the Security Council, the United States representative has made a strong appeal for a resolution which calls for the dismantling and withdrawal from Cuba of all nuclear missiles and offensive weapons and for international supervision of this process by a United Nations Observer Corps. The resolution also urgently recommends that the United States and the Soviet Union should confer promptly on measures to remove the existing threat to the security of the western

hemisphere and the peace of the world, and to report thereon to the Security Council ... As the House knows, Sir Patrick Dean, speaking on behalf of Her Majesty's government, gave his support to this resolution ... The British government are, of course, concerned that this new threat to security should be dealt with as rapidly as possible and will add their support to any measures which genuinely lead to that end. They trust also that, based upon some alleviation of the present state of tension, it might be possible to move into a wider field of negotiation.[110]

The Cabinet next met on Monday 29th, after the crisis had been substantially defused by Krushchev's agreement to withdraw the missiles. The Chiefs of Staff did not record discussions of the implications of the crisis during October 1962 and the only Cabinet meetings were as described above. The Parliamentary discussion of Macmillan's statement was curtailed by a demand by Black Rod that members attend the Queen's Speech; after the prorogation, the debate was not resumed. There is a perception in the 21st century that this was a week of near-panic in the UK with government and public assuming a nuclear war was imminent and preparing accordingly. In the USA '[t]here was some panic. People ran to grocery stores and took the bread and the milk and, you know, once the speech was given, we all said, 'Oh, my God.' And it was the scariest week of our lives.'[111] This is not reflected in the contemporary British press; the *Daily Mirror* reported on Thursday 25 October that sixth-formers at Midhurst Grammar School had held a two-day strike, including a ten-minute vigil at the local war memorial. Other public reactions listed included a 600-strong CND demonstration outside the American Embassy, student protests at Manchester, Hull, Leicester, Birmingham and Swansea, and that a company director was keeping his children home from school.[112]

On the 26th, after reporting the exchanges in the Commons, the *Daily Mirror* commented that '[t]he Cuban crisis has brought the world to the brink of war. Yet there is hope that the shock will inspire Kennedy and Krushchev to negotiate more realistically than they have done in the past.'[113] The Cuban missile crisis was less than two years after CND's Aldermaston march of 1960 and the associated mass demonstrations, but does not appear to have seized the public imagination, except perhaps in retrospect. In this sense, the lack of government intervention was successful in minimising the domestic impact of the crisis, but there is no hint in the record that this was a deliberate objective.

Labour Party in government

By the time of the general election in 1964, CND was virtually a spent force politically and although it remained an active presence in UK politics, unilateral disarmament was not to become a significant political issue until the cruise missile debates nearly twenty years later. The 1961 Labour Party conference rejected unilateral disarmament, although it supported 'an end to British nuclear testing, as well as no first use of nuclear weapons, and abandonment of the British deterrent'.[114] In 1964 the Labour manifesto stated that '[Polaris] will not be independent and it will not be British and it will not deter ... We shall propose the renegotiation of the Nassau agreement.'[115]

As had previous prime ministers, Wilson created a small Cabinet committee to consider nuclear policy (MISC 17) which met at Chequers in November, to establish the position on the nuclear deterrent. This discussion started: '[h]aving decided in principle to continue the Polaris programme, the major question which faced us ... was how to get rid of Macmillan's commitment to put the submarines into the [NATO Multi-Lateral Force] MLF'.[116]

The MLF had been an American concept for a force of nuclear-armed ships and submarines to be manned by mixed NATO crews under American command and control. The initial manning trial on board USS *Claude D Ricketts* was a success but the project was shelved; it was not politically tenable for NATO allies (especially the UK) because sole authority to launch lay with the USA.[117]

The MISC 17 meeting initially discussed the number of Polaris submarines already under construction and developed into a consideration of different mechanisms of command and control for this force under NATO and national arrangements:[118] '[i]nstead we would propose the establishment of an Atlantic Nuclear Force to which we would commit all our nuclear weapons ... [President] Johnson agreed to drop the MLF in favour of the ANF. Within a year the ANF had also sunk without trace, because nobody wanted it.'[119] Healey (whom I have just quoted) seems to have been surprised that

> our decision to keep a British nuclear deterrent was never contested; opposition focused on our agreement to host the American Polaris submarines at Holy Loch. I had never hesitated to justify our policy on nuclear weapons, and had explained NATO's nuclear strategy in detail in the House of Commons.[120]

Despite the manifesto commitment, Wilson's Labour government did not renegotiate the Polaris Sales Agreement and Britain built and deployed four of the planned five Polaris submarines. The *volte-face* over the manifesto commitment on Polaris was carefully wrapped into detailed discussion of the Atlantic Nuclear Force at every opportunity. To general agreement at a workshop of senior participants in British nuclear policy and historians in 2007, Michael Quinlan observed that: '[t]he ANF seemed to me to have a dual purpose. One was to fudge the Labour Party problem, and the other was to kill the NATO Multi-Lateral Force.'[121]

As Healey observed, Wilson was treated surprisingly gently by the press, even by the standards of the time. *The Observer* considered that '[b]y committing himself to the principle of an Atlantic Nuclear Force, Mr Wilson may have got rid of the albatross of the independent deterrent'.[122] In the run-up to the 1965 Defence White Paper, *The Guardian* asked

> [a]re Labour's first measures – completion of four Polaris submarines, the ANF ... consistent with its election pledges? Some critics have taken them as going in the opposite direction. Why build any Polaris submarines? The answer is that these four must be paid for anyway and that they can help to prepare for nuclear interdependence ... Is it not a reversal of Labour's pledge? No: it is a step towards the right objectives.[123]

This 'point of no return' argument was not in fact true; Healey had been briefed on his arrival at the MOD that only two SSBNs were under construction and that all of them could be converted to the hunter-killer role at no significant cost. At Wilson's request, Wilson, Healey and Walker had kept this from MISC 17. Healey subsequently explained that

[we] thought that in this uncertain world into which we were moving, a few Polaris submarines would be worth more than the same number of hunter-killers, both because they would give Britain more influence, particularly in Washington ... Moreover their running costs would be only £4 million a year – about two per cent of the defence budget.[124]

Defence policy and the withdrawal from east of Suez

The issue of maintaining the British position as a world power dominated the foreign and defence policy of both Macmillan's and Wilson's governments, even as they oversaw the retreat from east of Suez. Darwin describes this as 'an attempt, inevitably muddled and incoherent, to come to terms with a further contraction of British world power, an attempt however to stabilise Britain's world position to retain its basic elements, not to abandon it altogether'.[125] In this, possession of nuclear weapons was perceived as a potential surrogate for empire; nuclear ownership was regarded by successive governments as a pre-requisite for remaining in the front rank of world powers: '[t]he issue is not Europe v East of Suez, the problem is whether we are an island off the north-west coast of Europe or a world power'.[126]

British military commitments in the Middle and Far East were almost continuously reviewed during the financially straitened 1960s, and nuclear capabilities were not considered as distinct or significant factors. They were simply aspects of existing commitments to be assessed alongside other political and strategic issues. The Central Treaty Organisation (CENTO) was established in a treaty signed in 1955 by Iran, Iraq, Pakistan, Turkey and the UK, designed to minimise Soviet influence in the Middle East. 'The Akrotiri base [Cyprus] is used for the four Canberra strike squadrons which we have declared to CENTO: they are the only CENTO forces with nuclear capability and virtually the only forces declared to CENTO other than those of its regional members'.[127]

There was no strategic imperative for nuclear weapons specifically in order to defend British interests east of Suez, and 'to suggest any change in our plans now might hamper the development of proposals for the control of nuclear forces in Europe'.[128] By 1968, having told President Johnson '[a]t the root of this is a still rather confused groping for the real role that Britain ought to be playing in the world',[129] Wilson's government came down firmly on the assumption that '[o]ur standing in the world depended on the soundness of our economy, and not on a world-wide military presence'.[130] This clearly did not allude to the retention of nuclear weapons as status symbols. It did, however, set the tenor of subsequent public strategic discourse, looking towards NATO and the defence of Europe as the key defence role, and to 'soft power' for Britain's world role.

The 1970s – Continuous at-sea deterrence

Much public defence interest during the late 1960s and early 1970s focused on the war in Vietnam, with the rump CND developing and using many of the tactics of civil disobedience and non-violent direct action that had been evolved through the anti-nuclear demonstrations of the 1958–63 period. The Polaris submarines were duly constructed and deployed, with continuous at-sea deterrence being established on 30 April 1969 when HMS *Resolution* relinquished 'ready' status to HMS *Renown*. Since then, at least one SSBN has been perpetually 'ready'. During the Cold War, 'ready' meant the SSBN was at a few minutes notice to fire (Quick Reaction Alert (Nuclear)). This profile has been amended by successive defence reviews after the end of the Cold War and the notice to fire is now measured in days, but the ready SSBN remains in all respects prepared to launch within the notice deemed appropriate by the government.

As predicted, missile technology rendered a purely air-delivered deterrent critically vulnerable to pre-emptive strike (attack on the ground), and the emergence of anti-ballistic missile (ABM) defences threatened to render the British Polaris system incapable of guaranteeing to destroy Moscow's leadership centres in a unilateral strike. In extreme secrecy, successive governments since Wilson's maintained Project Super Antelope (later Chevaline), which was designed to enhance the payload delivered by British Polaris missiles in order to overcome this shortcoming. Such discussions had to remain out of the public eye (indeed, as usual, they were kept to the very smallest sub-committee within the Cabinet) because to admit that such a development was necessary would call into question the credibility of the deterrent.[131]

The 1979 Labour Party general election manifesto stated:

> In 1974, we renounced any intention of moving towards the production of a new generation of nuclear weapons or a successor to the Polaris nuclear force; we reiterate our belief that this is the best course for Britain. But many great issues affecting our allies and the world are involved, and a new round of strategic arms limitation negotiations will soon begin. We think it is essential that there must be a full and informed debate about these issues in the country before the necessary decision is taken.[132]

There was an element of sophistry about this position. In particular, as Lord Owen argues, 'Callaghan wanted to be more open about issues, in '78 I think, and the Cabinet Secretary argued him out of that position'.[133] Discussion within the Cabinet was being pushed by the need to make a decision on a replacement for Polaris.

Almost below the level of public perception NATO had adopted various defence and deterrence postures since its inception: '[t]he 1970s were a sleepy time at NATO'.[134] In 1968, MC-14/3 'Overall strategic concept for the defense of the North Atlantic Treaty Organization Area', was agreed and remained the current strategy for most of the 1970s. This deterrence posture was predicated on a strategy to fight and win a war involving the use of 'tactical' nuclear weapons in defence of Europe. Such a credible defence strategy was seen as a deterrent in its own right and was predicated on a close link between the tactical nuclear forces deployed in Europe and the strategic nuclear forces of the USA. In the mid-to-late 1970s, decisions would need to be taken to renew these tactical nuclear forces; mostly centred around the use of aircraft with dual capability.[135] For the UK these involved the RAF Jaguar and a free-fall nuclear bomb the WE-177, although the Vulcan V-bomber remained in service with a nuclear role until 1981.

In 1977, Callaghan established the by-now-customary small ministerial Nuclear Policy Group comprising the Foreign Secretary

(Owen), the Defence Secretary (Mulley) and the Chancellor (Healey), to consider the options for replacement of Polaris. In October, that group reached 'general agreement on the desirability of maintaining an independent UK nuclear deterrent but the view was strongly expressed that the criterion on which the effectiveness of our existing deterrent was judged – namely its capacity to penetrate the ABM defences around Moscow and destroy forty per cent of the Moscow region – should be re-examined'.[136] In a prescient move, they also agreed that

> [w]e shall need to take decisions on the renewal or replacement of the British [tactical nuclear forces] weapons in the next year or two since their present life is limited to the early 1980s. These decisions will be influenced not only by the Alliance's study on the modernisation of TNF but will also take account of national factors such as the political implications of embarking on a replacement programme, costs and availability of scientific resources ... Ministers will not need to take decisions for another year or two.[137]

Faced with the latter recommendation, ministers decided not to take decisions for another year or two.

In November 1977, the Cabinet Secretary advised Callaghan that it would be necessary to consider key issues for a replacement for Polaris:

(a) The politico-military requirement (for what purposes would we want to have the system?)
(b) The main criteria (given the purposes, what must the system be able to do?) (The criteria for deterrence are already being studied by a group set up, in accordance with Ministers' instructions, to examine the continuing validity of the Moscow criterion for the effectiveness of the British deterrent.)
(c) The main characteristics (how best to do the task, taking account eg of technical and resource factors.)
Experience suggests that studies and decisions on these basic issues could take one or two years.[138]

Two groups were set up by the ministerial group, one, headed by the Chief Scientific Adviser to the MOD, Ronald Mason, looking at the Moscow criterion and the other, headed by Anthony Duff, looking at technical and system options.

Owen felt strongly that a like-for-like replacement was not necessary and, with the help of a small group within his private office he produced an extensive (classified) document for Callaghan's Nuclear Policy Group arguing for a different strategic nuclear deterrent – one based on submarine-launched cruise missiles. This challenged the assumptions and recommendations of the Cabinet Office studies and Owen felt that he could not have such a document attributed purely to the Foreign and Commonwealth Office (FCO) because '[Cabinet] didn't agree with me and anyhow they were all hung up on all that sorts of thing. I said to him [Callaghan] that I had largely put up this in order to have a proper debate. He was very encouraging.'[139]

There was an element of very careful management about the government position during this time; the terms of reference for the study groups stated:

> No decision on the future of the deterrent is needed during the lifetime of the present Parliament. The purpose of this study is to examine and report on all the factors which the next Government, of whichever political Party, will need to take into account when reaching that decision.[140]

Two months later, the Defence Secretary Mulley (one of only three ministers who knew of the Duff and Mason studies), answered Parliamentary Questions on studies into missile systems: '[w]e have no plans to develop a cruise missile or a successor to Polaris ... I have made it clear many times that we stand by the commitment in our election manifesto not to proceed with a new generation of nuclear strategic missiles'.[141] Mulley's language was very carefully negotiated in order to answer the questions without appearing to admit to a contravention of the manifesto commitment.[142] In April 1979 Hennessy reported in *The Times* a 'private but outspoken dispute with an all-party select committee of the Commons by [Mulley's] refusal to permit serving officers, civil servants and government scientists to give evidence about the options for a third generation British nuclear deterrent to replace the Royal Navy's Polaris submarine squadron'.[143]

This tight control of the knowledge of the existence of the Duff and Mason studies had no basis in security, merely in the presentational difficulties these studies would have caused for the Labour government. Callaghan's government laid competent groundwork for the continuation of the British strategic nuclear deterrent. This was completed, in violation of the spirit, if not the letter, of the manifesto commitments on which the government had been elected. In particular, the commissioning of the Duff and Mason studies was a significant move 'towards the production of a new generation of nuclear weapons or a successor to the Polaris nuclear force' and the debates engendered by Owen's cruise missile papers tested many of the assumptions that informed the case for Trident; but the wider Cabinet, and the public, were perforce ignorant of these developments.

Similarly, in what could have been presented as a coup for British defence on a par with Macmillan's Polaris negotiations, during their meeting in Guadeloupe in January 1979 Callaghan had already agreed with President Carter in principle to the sale of the US Trident C4 system on similar terms, but he was unable to announce this during that Parliament. Owen was certain that, had Labour won the 1979 election, Callaghan would have moved to replace Polaris with Trident: 'he definitely wanted to keep the nuclear deterrent and he would have argued, as Prime Minister, that we should do Trident and he told me really that he felt that the weakness in my case was that ... we couldn't have a deterrent that was based on unproven technology'.[144]

Much like Wilson's government fifteen years earlier, Callaghan's government had been elected on a mandate not to evolve the next generation of the British strategic nuclear deterrent. Once in office both had identified pressing imperatives that required the reversal of this position. Wilson appeared able to do so relatively openly, although it required some 'terminological inexactitude' on the status of the Polaris submarines to achieve this. Callaghan's government maintained the strictest secrecy over the Duff and Mason studies, and deployed very carefully worded statements in public and Parliament in order to sustain the fig leaf of the 1974 manifesto commitment. The 1979 Labour manifesto opened the way for a successful Labour government to exploit the activities of Callaghan's Cabinet committees: '[w]e think it is essential that there must be a full and informed debate about these issues in the country before the necessary decision is taken'.

This linkage was upset by the Soviet deployment of SS20 short-range, nuclear-capable, ballistic missiles in eastern Europe in 1979; these could have reached European capitals, but not the mainland USA and therefore threatened to reduce the confidence of the Western allies that Soviet use of nuclear weapons in Europe would trigger the American guarantee to use their strategic forces; if the USSR could fight a nuclear war entirely without threatening the USA, would the USA invite retaliation against American cities by using its strategic forces to defend European capitals? The deployment of the SS20s would have significant repercussions.

Notes

1. HC Deb. 28 April 1937, *Hansard*, vol. 323 cols 312–9.
2. For a revisionist view of the importance of the attacks on the end of the Second World War, see W. Wilson, 'The myth of nuclear deterrence', *Nonproliferation Review*, 15 (2008), 421–39; W. Wilson, *Five myths about nuclear weapons* (Boston, MA: Houghton Mifflin Harcourt, 2013); and J. McMahan, *Killing in war* (Oxford: Oxford University Press, 2009).
3. C. Attlee, 'The atomic bomb: memorandum by the Prime Minister'. GEN 75/1, 28 August 1945. TNA CAB 130/3.
4. C. Attlee, GEN 75/3 'The atomic bomb: letter from the Prime Minister to President Truman', 25 September 1945. TNA CAB 130/3.
5. Admiralty 1945b, Director Plans, 2 September 1945. TNA ADM 1/117259.
6. Admiralty 1945a, ACNS(W), 15 August 1945. TNA ADM 1/117259.
7. HC Deb. 21 August 1945, Atomic Energy Committee, *Hansard*, vol. 413 cols 442–3.
8. Attlee, 'The atomic bomb: memorandum'.
9. The 'Groves–Anderson' memorandum (16 November 1945), reproduced in Gowing, *Independence and deterrence*, 85.
10. S. J. Anderson, Advisory Committee on Atomic Energy, 'Memorandum on international control of atomic energy', 5 October 1945. TNA CAB 130/3.

11 Cabinet Office. GEN 75/10, 'International control of atomic energy; report by officials', 29 October 1945. TNA CAB 130/3.
12 Cabinet Office. GEN 75/7th meeting: note of a meeting of ministers held at No. 10 Downing Street on Thursday 1 November 1945. TNA CAB 130/2.
13 Gowing, *Independence and deterrence*, 20.
14 Cabinet Office. DO(46) 17th meeting, Cabinet Defence Committee, minutes of a meeting held at No. 10 Downing Street on Monday 27 May 1946. TNA CAB 131/1.
15 Cabinet Office. DO(46) 5th meeting, Cabinet Defence Committee, minutes of a meeting held at No. 10 Downing Street on Friday 15 Feb 1946. TNA CAB 131/1. See also Alfred Goldberg, 'The atomic origins of the British nuclear deterrent', *International Affairs (Royal Institute of International Affairs 1944–)*, 40 (1964), 409–29: 427.
16 Williams, *Twilight of empire*, 118.
17 Note by the Controller of Production of Atomic Energy at the confidential annex to Cabinet Office. GEN 163/1st meeting, note of a meeting of ministers held at No. 10 Downing Street on Wednesday 8 January 1947. TNA CAB 130/16.
18 Brian Cathcart, *Test of greatness: Britain's struggle for the atomic bomb* (UK: Endeavour Press, 2016) [Kindle edn], location 519.
19 Chapman Pincher, *Into the atomic age* (London: Hutchinson & Co., 1948), 104.
20 'UK Government DSMA Notice 02: nuclear and non-nuclear weapons and equipment'. Defence and Security Media Advisory (DSMA). Available: www.dsma.uk/notice/nuclear-non-nuclear-weapon-systems-equipment/ [accessed 5 August 2021].
21 HL Deb. 17 March 1948, *Hansard*, vol. 154 cols 863–926.
22 HC Deb. 1 March 1948, Defence, *Hansard*, vol.448 cols 87–160: 87.
23 *Ibid.*, cols 109–10.
24 HC Deb. 12 May 1948, *Hansard*, vol. 450 col. 2117.
25 Nicholas Wilkinson, *Secrecy and the media: the official history of the United Kingdom's D-notice system* (Abingdon: Routledge, 2009), 223.
26 P. T. Siemes, 'The atomic bomb on Hiroshima: an eye-witness account', *The Irish Monthly*, 74 (1946), 93–104: 93 and 'The atomic bomb on Hiroshima: an eye-witness account (continued)', *The Irish Monthly*, 74 (1946), 148–54: 151.
27 Patrick B. Sharp, 'From yellow peril to Japanese wasteland: John Hersey's *Hiroshima*', *Twentieth Century Literature*, 46 (2000), 434–52: 434.

28 *The Guardian*, 'Hiroshima' (3 September 1946), p. 4 col. 1.
29 *The Observer*, 'Travelling third – marathon', (20 October 1946), p. 2.
30 *The Observer*, 'The atom on the air' (10 November 1946), p. 2.
31 *The Times*, 'Atomic bombs' (26 September 1946).
32 The history of how and why this cooperation was withdrawn, and then reinstated nine years later, is not directly pertinent here. The relevant implications will be considered, but the detail is very comprehensively addressed in John Baylis, *Ambiguity and deterrence: British nuclear strategy 1945–1964* (Oxford: Oxford University Press, 1995) and Gowing, *Independence and deterrence*.
33 Baylis, *Ambiguity and deterrence*, ch. 2.
34 W. Penney, unreferenced handwritten notes to Portal on high explosive research project 1946–7. TNA AB16/1905.
35 Cathcart, *Test of greatness*, location 2594.
36 Cabinet Office. Minutes of a meeting held at 10 Downing Street on Thursday 24 April 1952. TNA ADM 116/6087.
37 Air Staff Operational Requirement 229. TNA AIR20/2240.
38 Alfred Goldberg, 'The military origins of the British nuclear deterrent', *International Affairs (Royal Institute of International Affairs 1944–)*, 40 (1964), 600–618: 606.
39 Cabinet Office. Note of a meeting in the Cabinet Office, GEN 465, 12 March 1954. TNA CAB 130/101.
40 Cabinet Office. Minutes of a meeting of the Cabinet, 22 March 1954. TNA CAB 128/27/21.
41 A hydrogen bomb uses nuclear fusion, rather than the fission used by the atomic bomb. This releases considerably more energy, and uses a fission device simply as the 'primer' for the fusion reaction. The first fusion explosion had been a test of a particular design in 1952, but the Castle Bravo test was a fully weaponised device, ready to use as a bomb.
42 *Daily Mirror*, 'The monster' (2 April 1954), pp. 7–8: 8.
43 Richard Crossman, *The backbench diaries of Richard Crossman* (London: Hamish Hamilton and Jonathan Cape, 1981), 303.
44 Harold Macmillan and Peter Catterall, *The Macmillan diaries; the Cabinet years 1950– 1957* (London: Pan, 2012), 302.
45 *Daily Mirror*, 'Churchill confesses' (31 March 1954), p. 1.
46 Macmillan and Catterall, *Macmillan diaries, Cabinet years*, 305.
47 Cabinet Office. 'Memorandum by the Chiefs of Staff for the Defence Policy Committee and Cabinet: United Kingdom defence policy', 31 May 1954. TNA CAB 129/69. Paras 3, 5 and 21.
48 Macmillan and Catterall, *Macmillan diaries, Cabinet years*, 327.

49 Cabinet Office. Minutes of a meeting of the Cabinet, 8 July 1954. TNA CAB 128/27/48.
50 Macmillan and Catterall, *Macmillan diaries, Cabinet years*, 328.
51 Cabinet Office. Minutes of a meeting of the Cabinet, 8 July 1954. TNA CAB 128/27/48.
52 Norman Brook's handwritten notes of Cabinet, 8 July 1954. TNA CAB 128/27/48.
53 Macmillan and Catterall, *Macmillan diaries, Cabinet years*, 297.
54 Cabinet Office. 'Memorandum: United Kingdom Defence Policy', para. 23.
55 *Ibid.*, para. 32.
56 Cabinet Office. Cabinet report by the Committee on Defence Policy, 27 July 1954. TNA CAB 129–69–0050.
57 *Ibid.*, para. 19.
58 *Ibid.*, Annex II.
59 MOD Memorandum, Macmillan (MOD) to Churchill (Prime Minister), 8 December 1954. TNA DEFE 13/45.
60 Cabinet Office 1955. D(55)17 'The defence implications of fall out from a hydrogen bomb' (the Strath report). TNA CAB 21–4054.
61 A. Anstey, SG(60)35, 'Note on the concept and definitions of breakdown', 10 June 1960. TNA DEFE 10/402.
62 See: Matthew Grant, *After the bomb: civil defence and nuclear war in Britain 1945–68* (London: Palgrave Macmillan, 2009).
63 Cabinet Office. Unreferenced memorandum, Monckton (Cabinet Office) to Chilvers (MOD), 11 June 1956. TNA CAB 21–4054.
64 Taylor and Pritchard, *The protest makers*, 2 and Crossman, *Backbench diaries*, 387.
65 Taylor and Pritchard, *The protest makers*, 4.
66 Benn, *The Benn diaries*, 4.
67 Aneurin Bevan, 'Labour Party Conference speech, Brighton, 3 October 1957' in Brian MacArthur, *The Penguin Book of Twentieth-Century Speeches* (London: Penguin, 1994).
68 Mark L. Harrison, 'CND: the challenge of the post-Cold-War era' (1994), PhD thesis, University of Loughborough.
69 G. F. Kennan, Reith Lecture, 'Russia, the atom and the West', [radio]. London: BBC. 1957.
70 *The Times*, 'Behind the voices' (8 March 1958), p. 7.
71 *The Times*, 'Nuclear tests challenge' (1 April 1958).
72 *The Times* (Monday, 17 February 1958), p. 9 col. A.
73 Taylor and Pritchard, *The protest makers*, 6.
74 *Ibid.*, 9.

75 *The Times* (5 April 1958), p. 6 col. F.
76 *Daily Mirror*, 'Only 11 miles to go now' (7 April 1958), p. 8.
77 Taylor and Pritchard, *The protest makers*, 12.
78 'Freedom 16 April 1960' quoted in Paul Mercer, *'Peace' of the dead: the truth behind the nuclear disarmers* (London: Policy Research Publications), 70.
79 *Ibid.*
80 Philip Williams, *Hugh Gaitskell: a political biography* (London: Jonathan Cape, 1979), 369.
81 Crossman, *Backbench diaries*, 883.
82 Gaitskell's position was reasserted to such an extent that even Wilson's candidacy as a moderate able to bring the party together was derided: '[i]f the Labour party ends this week facing in two directions, it is certain that the figure of Mr Wilson will be there, at the end of both of them' (*New Left Review*). Quoted in Williams, *Hugh Gaitskell*, 372.
83 MOD BND(SG)(59)1; British Nuclear Deterrent Study Group – minutes of a meeting held Monday 16 July 1959. In: MOD (ed.), TNA DEFE 10/665.
84 A ballistic missile is launched into space and 'falls' back to earth at speeds around Mach7. A cruise missile flies to its target rather like an unmanned aircraft, and is similarly vulnerable to interception. Ballistic missiles have ranges around 10,000km but cruise missiles, even in the 2020s, are limited to around 2,000km.
85 Three jet bomber types replaced the Second World War bomber fleet during the 1950s: the Vickers Valiant, the Avro Vulcan and the Handley-Page Victor, known as the V-bombers.
86 Cabinet Office. BND(SG)(59)15 'British Nuclear Deterrent Study Group – draft interim report', 30 October 1959. TNA DEFE 10/665.
87 Cabinet Office. BND(SG)(60)3, 12 April 1960. TNA DEFE 10/665.
88 To put this into perspective, the Minot air force base in North Dakota houses one of three US intercontinental ballistic missile (ICBM) wings (each wing operates about 150 ICBM silos) and covers a territory slightly larger than Wales.
89 The Memorandum of Understanding was signed on 6 June 1960: Cabinet Office C(60)97, 'Skybolt – note by the Minister of Defence', 20 June 60. TNA CAB 129/101.
90 Ken Young, 'The Skybolt crisis of 1962: muddle or mischief?', *Journal of Strategic Studies*, 27 (2004), 614–35: 614.
91 MOD (Air Staff), unreferenced note, Sec. BNSG to Sir Solly Zuckerman, 'Future deterrent policy', 26 September 1960. TNA DEFE 19/11.

92 MOD, unreferenced memorandum, Lord Selkirk to Lord Hailsham, 1 January 1958. TNA/ ADM/1/27375.
93 Peter Hennessy and James Jinks, *The silent deep: the Royal Navy submarine service since 1945* (UK: Penguin Random House, 2015), 202.
94 Cabinet Office. Conclusions of a meeting of the Cabinet, Thursday, 28 July 1960. TNA CAB/128/34.
95 Cabinet Office. Conclusions of a meeting of the Cabinet held at Admiralty House on Thursday 15 September 1960. TNA CAB/128/34.
96 Ibid.
97 Harold Macmillan and Peter Catterall, *The Macmillan diaries; Prime Minister and after* (London: Pan, 2011), 526.
98 Ibid., 528.
99 Principally the *Daily Express*, which then had very high circulation; also (in London) the *Evening Standard*.
100 Macmillan and Catterall, *Macmillan diaries, Prime Minister and after*, 527.
101 Ibid., 508.
102 *Daily Express*, '1am: blockade on' (23 October 1962), p. 1.
103 *The Guardian*, 'Soviet deception on Cuba' (23 October 1962), pp. 15., 1 and 8.
104 Cabinet Office. CC(62), 61st conclusions of a meeting of the Cabinet, 23 October 1962. TNA CAB/128/36.
105 Ibid.
106 Quoted verbatim in *The Guardian*, 'Britain behind Kennedy: "Soviet guilt"' (24 October 1962), p. 1.
107 *Daily Mirror*, 'US hunts target no. 1 – the red rocket runner' (24 October 1962), pp. 1 and 32.
108 *Daily Mirror*, 'Courage – but is it wise?' (24 October 1962), p. 2.
109 Cabinet Office. CC(62), 62nd conclusions of a meeting of the Cabinet, 25 October 1962. TNA CAB/128/36.
110 HC Deb. 25 October 1962, *Hansard*, vol. 664 cols 1053–64: 1054.
111 CBS News, 'Remembering Cuban missile crisis, 50 years later', CBS News (2012). Available: www.cbsnews.com/news/remembering-cuban-missile-crisis-50-years-later/ [accessed 17 December 2012].
112 *Daily Mirror*, 'No-war strike in the sixth form' (25 October 1962), p. 32.
113 *Daily Mirror*, 'Helpful crisis?' (26 October 1962), p. 2.
114 Len Scott, 'Labour and the bomb: the first 80 years', *International Affairs*, 82 (2006), 685–700: 689.

115 'The new Britain', Labour Party (1964). Available: http://labourmanifesto.com/1964/1964-labour-manifesto.shtml [accessed 23 September 2015].
116 Denis Healey, *The time of my life* (London: Michael Joseph, 1989), 304.
117 Bastiaan Bouwman, 'Present at the undoing: the Netherlands and the multilateral force', Nuclear Proliferation International History Project (2013). Available: www.wilsoncenter.org/publication/present-the-undoing-the-netherlands-and-the-multilateral-force [accessed 27 July 2017].
118 Cabinet Office. MISC 17/4, Cabinet defence policy, minutes of a meeting at Chequers, 22 November 1964. TNA CAB 130/213.
119 Healey, *The time of my life*, 305.
120 *Ibid.*, 351.
121 Peter Hennessy, 'Cabinets and the Bomb' workshop [online]. (London: British Academy, 2007). Available: www.britac.ac.uk/node/4986/ [accessed 27 July 2017].
122 *The Observer*, 'Comments: the deterrent' (20 December 1964), p. 7.
123 *The Guardian*, 'Towards a fresh defence policy' (22 February 1965), p. 10.
124 Healey, *The time of my life*, 302.
125 John Darwin, *Britain and decolonisation: the retreat from empire in the post war world* (London: Macmillan, 1988), 288.
126 Cabinet Office. Defence review: minute by Mr Walker to Prime Minister on the issue 'whether we are an island off the north west corner of Europe, or a World Power', 23 November 1965. TNA PREM 13/216 ff. 6–9, para. 2.
127 Cabinet Office. MISC 17/4, Defence review: report to ministers by an official committee (chairman Sir B. Trend) of the Cabinet Defence and Overseas Policy Committee, 8 November 1965. TNA CAB 130/213. Para. 37.
128 Ministry of Defence, 'The United Kingdom defence review; draft *aide mémoire* by HMG for discussion in Washington and Canberra', January 1966. TNA DEFE 13/477. Para. 6.
129 Cabinet Office. Foreign Office telegram no. 554: text of Mr Wilson's reply to President Johnson's letter of 11 January 1968, 15 January 1968. TNA PREM 13/1999.
130 Cabinet Office. CC(68)3 'Public expenditure: post-devaluation measures: Cabinet conclusions on withdrawal from east of Suez', 4 January 1968. TNA CAB 128/43, p. 6.
131 This is a fascinating discussion, based on the premise that in order to deter the USSR the UK had to be able to threaten to destroy Moscow

with a unilateral strike. This 'Moscow criterion' was the benchmark for the credibility of the deterrent throughout the life of the Polaris system, and argument about its continued relevance coloured both Chevaline development and discussion about the replacement for Polaris. Though the officer commanding a Vanguard class submarine, I knew nothing of the detail. See: Kristan Stoddart, 'Maintaining the "Moscow criterion": British strategic nuclear targeting 1974–1979', *Journal of Strategic Studies*, 31 (2008), 897–924.
132 'The Labour way is the better way', Labour Party (1979). Available: www.politicsresources.net/area/uk/man/lab79.htm [accessed 23 September 2015].
133 Lord David Owen, interview with Andrew Corbett, 16 April 2015.
134 Jamie Shea, '1979: the Soviet Union deploys its SS20 missiles and NATO responds', NATO (2009). Available: www.nato.int/cps/en/natohq/opinions_139274.htm [accessed July 2013].
135 'Dual-capable' aircraft are 'tasked and configured' to perform either conventional or theatre nuclear missions.
136 Cabinet Office. Conclusions of a ministerial meeting held at No. 10 Downing Street on Friday 28 October 1977 at 0945, TNA PREM16/1564.
137 Cabinet Office. A05828, 'Military nuclear issues', 25 October 1977. TNA PREM 16/1564.
138 Cabinet Office. A06085, 'Nuclear matters', 28 November 1977. Loose minute from Cabinet Secretary to PM. TNA PREM16/1564.
139 Owen, interview with Corbett.
140 MOD, memorandum from Parliamentary Private Secretary to Prime Minister to Cabinet Secretary, 'UK nuclear deterrent', 30 January 1978. TNA PREM 16/1654.
141 HC Deb. 21 March 1978, *Hansard*, vol. 946 cols 1313–15.
142 MOD, Parliamentary question handling brief, 21 March 1978. TNA PREM 16/1564.
143 *The Times*, 'Mr Mulley angers specialists over successor to Polaris' (30 April 1979), p. 1.
144 Owen, interview with Corbett.

5

The Polaris replacement decision

Mrs Thatcher's government was elected in May 1979 with a manifesto pledge to make significant increases in the level of defence spending:

> During the past five years the military threat to the West has grown steadily as the Communist bloc has established virtual parity in strategic nuclear weapons and a substantial superiority in conventional weapons ... The SALT [Strategic Arms Limitations Talks] discussions increase the importance of ensuring the continuing effectiveness of Britain's nuclear deterrent.[1]

As described in chapter four, the construction of ABM defences around Moscow had caused a significant concern about the credibility of the UK strategic nuclear deterrent as a unilateral deterrent. The Conservative government decision on 'ensuring the continuing effectiveness' of Britain's nuclear deterrent was substantially informed by the Duff-Mason Report which, exceptionally, Callaghan had instructed to be handed to the incoming Prime Minister. Mrs Thatcher also received private assurances from President Carter that he would honour the arrangement he had come to with Callaghan in which the USA would be prepared to negotiate the sale of Trident C4 missiles with MIRV technology (multiple, independently targetable re-entry vehicle), but Carter wished this kept private until SALT II was ratified. The decision to replace Polaris was managed by yet another small group of ministers (MISC 7), once more in extreme secrecy from both the public and the remainder of the Cabinet; in his initial briefing on the handling of nuclear deterrence matters, the Cabinet Secretary asked

Mrs Thatcher whether she 'propose[d] to confine it to yourself and the three Departmental Ministers directly concerned?'[2] She did.

Simultaneously, the government was establishing its position on the Long-Range Tactical Nuclear Forces and the SALT II treaties which were key to NATO's evolving nuclear strategy. The SALT discussions were bilateral negotiations between the USSR and the USA on nuclear arms control measures. The first round had led to the 1972 agreement on ABM defences and the Interim Agreement on strategic nuclear arms, and negotiations had immediately started on SALT II. Although the negotiations were bilateral, the UK's interests were mostly in ensuring that the agreement neither included British weapons in the totals, nor inhibited transfer of nuclear technology from the USA to the UK. There was continuous pressure from the USSR to include European systems, including the UK's nuclear forces, in the USA's 'count'.[3]

Early advice to Francis Pym (Secretary of State for Defence) on the Long-Range Tactical Nuclear Forces treaty suggested that:

> a deployment of [ground-launched cruise missiles (GLCM)] in the UK could well offer an additional focus for demonstrations by CND-style and perhaps environmentalist critics. This likelihood, and the desirability of keeping its impact to a minimum, must feature among the factors affecting the selection of sites and planning of dispersal patterns and any off-base exercises.[4]

Pym seems not to have taken this advice to heart; '[h]is reaction was not quite that he saw no difficulty, but that he saw none of an order which should be allowed to deflect us from a GLCM deployment if that was otherwise clearly the right course on the grounds of national security'.[5] Pym was convinced that the public would need to be educated about the case for a replacement for Polaris, but his successes were limited and the challenges manifold.

Cabinet Office preparations for Mrs Thatcher's first MISC 7 on 24 May 1979 emphasised on one hand the absolute imperative for secrecy, and on the other the need to consult more widely in order to further inform the decision. The initial briefing on the nuclear deterrent by the Cabinet Secretary (John Hunt) to the Prime Minister outlined the history of the Polaris force, its vulnerability to the evolving ABM systems, and the Chevaline project including its £935 million cost. The briefing considered the ongoing UK/US collaboration on the Polaris project and predicted that the missiles, systems and submarines would remain sound: 'it should therefore be possible to maintain the present force in operation until the mid-1990s, albeit with increasing costs and technical problems ... If we are to develop a successor system, it will need to enter into service by the mid-1990s.'[6]

In his follow-up briefing, Hunt elaborated the key themes of the Duff-Mason Report and highlighted the key decisions necessary. He also observed that the

> general approach is on orthodox lines, but it represents an attempt, for the first time in recent years, to work out a <u>concept for the United Kingdom deterrent</u> ... the absence of any conclusions is deliberate; the intention was to provide Ministers with arguments on either side on the basis of which they could reach a decision in principle.[7]

This clearly suggests that the Cabinet Office viewed the key decisions about Polaris replacement as not technical in nature, but intensely political. Hunt then explicitly linked the technical with the political: 'what constitutes unacceptable damage and thus what would deter attack on the United Kingdom are essentially matters of political judgement'. He described the close linkage between the targeting requirement (Chevaline was procured to meet the perceived need to threaten Moscow),[8] necessary performance criteria and therefore cost: '[t]he cost of continuing to provide this capability in a successor system is likely to be very high'.[9]

It is in Hunt's paper that the close link between highly classified nuclear targeting considerations and public statements about policy and system procurement are first highlighted for the Conservative government. Discussion of the Polaris successor continued throughout the autumn in MISC 7, with the main issues being technical. Although there was no debate in the committee about whether the UK required a nuclear deterrent, there was a very pertinent exchange debate between Mrs Thatcher and Hunt's replacement as Cabinet Secretary, Robert Armstrong, in her preparations for a November meeting of MISC 7. In his initial, preparatory briefing for this meeting, Armstrong had written to the Prime Minister '[i]n considering this report I believe that the Ministerial Group will wish to concentrate on three questions. These are:– (a) Should there be a British strategic nuclear deterrent in succession to Polaris? ...' Mrs Thatcher noted by hand on the brief '[w]e have decided there should'.[10]

Despite this definitive clarification, Armstrong's subsequent briefing on suggested conduct of the meeting returns to the issue repeatedly, suggesting he views it as an important aspect of the decision; and this correspondence is worth considering in substance:

> This is a <u>key meeting</u>. The likely decisions will affect our most important means of defence over the next 40 years and thereby the basis of our international military posture ...
>
> The Chancellor of the Exchequer and the Secretary of State for Defence have (I understand) reached effective agreement on the formula proposed by Sir John Hunt ... under which the Defence Budget ... to 1983–83 is fixed at the (lower) Treasury figures and the cost of replacing Polaris is treated as a charge on the Contingency reserve. This will not have been reported to the Cabinet by 5th November.[11]

Armstrong draws attention to the financial manipulation that was required to obscure from Cabinet the necessary 'long lead' financial commitments for the Polaris successor. Such 'economy with the truth'[12] was not new; after all, Mrs Thatcher's government had inherited the top secret Chevaline programme and its huge (and as yet unpublished) budget, and the suggested use of the contingency reserve would perpetuate this 'economy' within Cabinet and, by extension, Parliament that had characterised British nuclear-policy decision-making since Attlee.

> 3. You may wish to conduct the meeting in three stages:
>   (i) Procedural points
>   (ii) The answers to the three questions in my minute of 29th October –
>   (a) Do we retain our strategic deterrent?
>   (b) What should it be capable of doing?
>   (c) Which weapon should we choose? ....
>
> 5. Your Luxembourg speech made pretty clear that we could continue with our deterrent after Polaris; and it was MISC 7's starting point, at its first meeting in May, that the Government was fully committed to doing so. But the Chancellor was not invited

to that meeting; and the seriousness of the issue is such that your colleagues should at least be invited to reaffirm that we do wish to stay in what is, for us, a pretty big league. Mr Pym could be invited to begin, on the basis of his paper.

6. That paper deals summarily with the question, on the basis of previous discussion. Are you content to do that? Or do you want to invite the Committee to go over the fundamental questions again? [Mrs Thatcher hand-wrote 'NO' against this question.] I suppose these questions are: What good has it done us so far to be in the strategic deterrent league? Given the decline in our world position in other respects, will it do us enough good to stay in the league from the 1990s to justify the cost of the burden this programme represents in the meantime? How important is it for us, and for our NATO Allies, that we should continue to maintain our own strategic deterrent capacity? How conceivable is it that we may want to use, or to be able to threaten to use, a British strategic deterrent independently of the United States, either in our national interest or in that of the NATO Alliance? In terms of cost-effectiveness as a contribution to our own and our Allies' security, is this preferable to use of a corresponding amount of resources on more conventional weapons? ... I suspect that answer to all these questions is in effect that, having been in the league for thirty years, it is inconceivable for a Government committed as this is to the maintenance of national defence to take a decision which would irrevocably take us out of that league. But it may still be right to have asked, and agreed upon answers to, questions of this kind.[13]

Paragraphs three, five and six readdress his basic question – should there be a replacement? Armstrong outlines a series of not unreasonable questions and concludes paragraph six with the point that, even if the answers to these are not in fact central to the decision to replace Polaris, the government would do well to have articulated the factors and considered how to respond when others (inevitably) asked the questions. These questions would be regularly reflected in the increasingly polarised discourse on nuclear deterrence for the remainder of the Parliament.

By early November, MISC 7 discussion was focused on the relative merits of submarine-launched ballistic missile (SLBM) systems and submarine-launched cruise missile (SLCM) systems. At the September meeting air-launched cruise missiles had been ruled

The Polaris replacement decision 103

out as an option for a strategic nuclear deterrent.[14] The decision to procure Trident C4 was taken in the delayed MISC 7 meeting in December, although the minutes of that meeting remain classified and it is not clear whether Armstrong's fundamental questions were addressed. They certainly appear to have played no part in government preparation for the subsequent Parliamentary scrutiny processes, nor the public discourse. *The Guardian* picked precisely on Armstrong's unanswered questions, speculating:

> if Mrs Thatcher and her Ministers do endorse the Trident plan next week, its public debate is likely to be vigorous … The Government is bound to be challenged as to why Britain still needs an 'independent' nuclear deterrent, why the Polaris system cannot be modernised and, above all, why the deterrent force should not consist of much cheaper nuclear-armed cruise missiles of the kind we are in any case proposing to have based in this country … as part of the collective plan to modernise NATO's so-called theatre nuclear weapons.[15]

The same article stated that a 'decision has been taken at the Ministry of Defence that Britain's independent nuclear deterrent should be replaced by a fleet of five submarines carrying American Trident missiles fitted with British warheads. A recommendation will be submitted next week to a special Cabinet sub-committee, chaired by Mrs Thatcher, which is expected to endorse the plan.' Outside MISC 7, extreme secrecy was being exercised until a decision had been made and publicity became inevitable. Armstrong briefed the Prime Minister on efforts to stem the leak, assessing that there was little an able defence correspondent could not have picked up, but '[w]hat is new, and in the light of President Carter's message extremely damaging, is the reference to the fact that the subject is shortly to go to a Committee of Ministers'.[16] Negotiations with the US authorities had been ongoing throughout the autumn and Robert Wade-Gery of the Cabinet Office had led a small team of officials to Washington to agree the outline of the deal with President Carter's deputy National Security Advisor, Dr David Aaron.

The guidance on the publicity to be afforded this visit was classified Top Secret UK Eyes Alpha. This was the same classification as the MISC 7 main working papers considering detailed UK nuclear policy and capabilities, SSBN patrol reports and Polaris targeting instructions. Top Secret is defined as the '... most sensitive information requiring the highest levels of protection from the most serious threats. For example where compromise could cause widespread loss of life or else threaten the security or economic wellbeing of the country or friendly nations.'[17] UK Eyes Alpha further limits access to UK nationals only. Although the policy papers have been declassified, the Polaris patrol reports and targeting instructions have not; although policy decisions become historic and their release does not compromise current policy, even historic operating patterns and philosophies remain classified, precisely because they do not lose their relevance for contemporary operations.

While the UK considered the Polaris successor in utmost secrecy, NATO was dealing with proposals to modernise the alliance's theatre nuclear weapons capabilities as a response to the Soviet deployment into eastern Europe of SS20 intermediate-range nuclear missiles. Despite considerably more public debate across Europe about the NATO strategy, the British government persisted in dealing with these issues as substantially unrelated to the strategic deterrent and the decision over a successor for Polaris.[18] This myopic insouciance is all the more surprising given the advice received from Michael Quinlan (MOD deputy under-secretary (policy)) in July and how sensitive to security aspects were the negotiations for Trident.

The NATO theatre nuclear forces issue

The Soviet Union began deploying medium-range SS20 ground-launched ballistic nuclear missiles in Eastern Europe between 1976 and 1978. SS20 were able to threaten all of Europe's capitals while not posing a threat to the USA, thus upsetting what was considered a regional and strategic balance by many in NATO. This posed the prospect of 'de-coupling' a threat to Europe from the threat to the USA, and thus reducing the likelihood of US strategic retaliation and the credibility of NATO's nuclear deterrence strategy.

The Soviet move was carefully calibrated to challenge the cohesion of the alliance on this highly emotive issue and was accompanied by an aggressive Soviet media campaign highlighting the withdrawal of 20,000 Soviet troops from eastern Europe and an offer to reduce tactical nuclear warheads in Europe, if no new NATO systems were deployed. The challenge to alliance cohesion was readily perceived; Pym advised Mrs Thatcher that '[t]he modernisation of long-range theatre nuclear forces is of high importance to NATO defence. It has moreover become of political significance reaching beyond the strictly defence considerations; it is now a key test of NATO's collective will to ensure its security.'[19] The increased visibility generated by the Soviet PR campaign re-ignited the anti-nuclear campaigns across Europe, with protests and membership rising rapidly between 1978 and 1982.

NATO's response options were being considered in the newly formed High Level Group (HLG). The alliance agreed a 'twin track' approach – one of the most sophisticated arms control measures of the Cold War – NATO linked deployment of similarly capable NATO systems (Tomahawk GLCM and Pershing MRBM) to the deployment of SS20, but offered to negotiate arms control measures with the USSR from the position that the NATO systems would be deployed unless the SS20 were removed. The NATO strategy led to the deployment of GLCM in the UK, the Netherlands, Belgium, Italy and West Germany, and Pershing II MRBM in West Germany, leading to intense opposition from anti-nuclear campaigners across Europe.[20] The US deputy National Security Advisor, Aaron, toured the capitals of the putative host nations (UK, Federal Republic of Germany, Holland, Belgium and Italy) lobbying for a robust response to the Soviet campaign. British officials advised that 'while public presentation was perhaps less of a problem in the UK than in some other countries, it nevertheless needed to be handled carefully and the question of the basing facilities was particularly sensitive'.[21]

The NATO HLG plan was considered by MISC 7 in early December and it was agreed that, prior to Mrs Thatcher's visit to Washington on 17 December, the Defence Secretary would make a statement to Cabinet on the outcome of the meeting on 12 December of the NATO Nuclear Planning Group (NPG) (the body to which the HLG reports). Armstrong suggested that this might be an opportune moment to inform Cabinet of the decision to procure Trident, but highlighted the American desire for continued discretion.[22] In the event, Pym reported the NPG outcome to Cabinet, but not the Trident decision. NATO announced the decision to adopt the 'twin track' stance on 12 December. US Secretary of State Cyrus Vance announced 'I believe that our governments can be proud of this memorable achievement and that the free people of the alliance will show overwhelming support for the decisions made here today'.[23] But Vance had misjudged European public opinion; according to one CND history, 'even as the decision was being announced, 40,000 people were gathering at the NATO headquarters in Brussels to protest; the antimissile movement turned out to be the greatest wave of protest that had taken place in western Europe since World War II'.[24]

The CND campaign selectively used information from Vance's speech, inaccurately (but very emotively and effectively) disparaging the Pershing II and GLCM as 'first strike' weapons. First strike is a strategic concept, envisaging one side having the ability to use sufficient precision weapons with enough confidence that they could successfully destroy an opponent's ability to retaliate, before the opponent could react; it is often referred to as a 'disarming first strike'. A credible first-strike capability has three requirements; the first-strike weapons must be accurate enough to target the opponent's nuclear weapons to a high degree of assurance – both Pershing and GLCM were capable of this degree of accuracy; there must be sufficient first-strike weapons to target all of the opponent's weapons simultaneously – the 464 GLCM and 108 Pershing II missiles announced could target around 10 per cent of the 5,000 Soviet nuclear weapons in Europe; and the opponent must not have an assured second-strike capability, such as submarine-launched ballistic missiles which are immune to first strike – the USSR had SSBNs in its Northern Fleet and based in the Pacific.[25]

The anti-nuclear campaigns ignored the elements of Vance's speech that did not suit this narrative:

> The modernization decision that we have made here also makes it possible for us to withdraw 1,000 nuclear warheads from Europe ... Thus, far from increasing NATO's reliance on nuclear weapons, our decisions will result in a significant reduction in the size of NATO's overall nuclear stockpile in Europe.[26]

NATO deployment of cruise missiles, at the same time as the decision was being taken to replace Polaris, was to prove a demanding period for British nuclear policy formulation.

Mrs Thatcher met with President Carter in December and verbally agreed the outline terms of the Trident purchase. The formal exchange of letters was further delayed at Carter's request in order to avoid ramifications for the SALT II treaty, which was about to be ratified – he was concerned that the USSR would seek further concessions if it was announced that the UK would deploy Trident. Specifically, he also requested that the proposals should not be put to the British Cabinet. When asked whether the negotiations on the exchange could continue in the meantime:

> Dr Brzezinski [US National Security Advisor] indicated that this would be negotiable, provided that discussion was confined to the same restricted group of people as had been involved hitherto. It would not, however, be possible for technical discussions to proceed without extending the circle of those involved, which the President did not want to do.[27]

This note was classified top secret, and the implicit ramifications of a leak for SALT II seem to vindicate that level of discretion, although it does say little for the perception of British security that Cabinet and the Chiefs of Staff were additionally explicitly excluded.[28] In spring 1980, there was a further delay to the announcement as the US administration sought to avoid perceptions of over-reaction to the Soviet invasion of Afghanistan.[29]

Pym had long advocated a more open public approach and debate in the Commons and an adjournment debate on nuclear defence issues had been scheduled for 24 January. He opened the debate by indicating that it was the first such since 1964 and that he would concentrate on nuclear issues and nuclear policy. He considered that 'the arguments surrounding nuclear strategy neither should be, nor can be, taken for granted. They require constant rethinking and restating, and I feel sure that it is right for the House to play its part in that process.'[30] Having opened with a brief précis of NATO strategy, in a 'world where nuclear weapons exist', against a 'potential adversary who has built up ... a vast – and offensively structured – apparatus of military power...' Pym described NATO's modernisation programme, in particular the need for a modern capability to strike the Soviet homeland from within Europe. He informed Parliament of the decision to base 160 American ground-launched cruise missiles in the UK, and sketched the arrangements for political control of these missiles. He covered the UK contribution to NATO deterrence, and then described the technical aspects of the systems involved.

Pym also described the Chevaline upgrade to the Polaris system and, for the first time, its £1 billion cost was publicly disclosed. He carefully avoided any reference to specific Anglo-American negotiations and agreements, referring only to the US commitment to cooperate in providing a UK strategic nuclear deterrent. In closing, he described the likely costs of replacing Polaris in language that suggested it was an inevitable expenditure: '[t]hat is, of course, still a massive demand on our limited resources, but we must keep it in the perspective of what modern defence inescapably costs'.[31] When faced with questions, he indicated that cruise missiles had been considered and rejected as strategic deterrence options, and that there was no fixed timetable for the decision on the Polaris replacement.

William Rodgers, the Labour defence spokesman, mocked the 'enthusiasm of the then Opposition for debating such matters and their lack of enthusiasm as displayed since May 1979'.[32] He then looked forward to the work of the Defence Select Committee in enabling the Commons to have a better-informed debate than would otherwise be the case. Rodgers queried '… whether it is wise to replace Polaris at all. To me, that remains an open question which should be subject to debate.'[33] The Labour MP John Cartwright followed this with:

> The Secretary of State for Defence put a rather different point of view on 18 December. He said, 'I want the greatest possible discussion about the matter.' … We are left with a gap in our information, and that cannot be filled in the time available to the new Select Committee on defence … The House will fail in its duty to the nation if the Government are not pressed to provide the basic information needed to decide whether their judgments are correct.[34]

The Times reported the exchanges under the headline 'more information must be given',[35] though this reflected more the tenor of the debate than its editorial opinion at that time. The *Daily Express* was more partisan: 'Maggie's Cold War – H weapons give boost to Polaris fleet',[36] enthusiastically reporting the increase in the capability of the Polaris fleet and the plan to position cruise missiles in the UK. *The Times* considered the subsequent House of Lords debate under the headline 'Implications of decision to buy Trident' and cited six Lords who argued against it, including Lord Carver (formerly Chief of Defence Staff).[37] Viscount Trenchard, the Minister of State for Defence, was quoted at length, which provided a balanced account of the Lords' debate, but with no editorial comment *per se*.

Inevitably, increased public interest and debate ensued, although they were substantially unencumbered by government participation. In public correspondence reminiscent of Spaight's during the Second World War, the recently retired Chief of Defence Staff Air Chief Marshal Cameron refuted arguments made against Trident by Lord Carver in letters to *The Times*,[38] but there was negligible government intervention in the public discourse.

CND had been virtually moribund since the mid-1960s. In the late 1970s anti-nuclear opposition experienced a resurgence in interest across Europe, in the USA and in the USSR because of NATO's discussions about tactical nuclear forces and the increasingly bellicose rhetoric of the USSR and USA. Regardless of official distinctions, NATO tactical nuclear forces and the successor to Polaris were regarded as a single issue by CND.

During the ministerial discussion about how to announce the decision to replace Polaris, which had run throughout June and July, there had been an extended debate on how much to reveal publicly about the decision itself, the nuclear deterrent and the Trident system. In keeping with his long-standing opinion, Pym advocated an official publication to set out as much of the argument and case for the successor as possible. The original draft of a public 'Open Government Document'[39] (substantially written by Michael Quinlan) was circulated to the Prime Minister, Home Secretary (William Whitelaw), Foreign Secretary (Peter Carrington) and Chancellor (Geoffrey Howe) on 10 June. Whitelaw responded: '[m]y only question is whether in these circumstances we are wise to expose as many of the details as you do. You may feel it is essential. I do, however, have the feeling that in this field it is wise to give <u>as little information as is possible</u>' (emphasis in original).[40] Carrington was generally content but suggested amending the text on the Nuclear non-Proliferation Treaty (NPT) in order to remain sensitive to the NPT Conference later in 1980,[41] and Howe sat on the fence: 'I am inclined to wonder whether it is really wise to say quite so much. We would expose a lot of flank. On the other hand, I recognise that Francis Pym is committed to publishing some account of the basis for our decision, and to say too little would be counter-productive.'[42] The Prime Minister agreed with Pym's intent, but also with the need to provide rather less information.[43]

Obviously the Labour opposition knew that a decision on a replacement for Polaris must be looming, even if they knew no details, and regularly pushed for a debate prior to any decision. On 10 July 1980, just prior to the summer recess of Parliament, the Leader of the House was repeatedly pressed to declare a date for a debate on the Polaris successor, to which he repeatedly responded that he had received no request for a debate on nuclear policy.

Ideally, the government wished to time the announcement to enable formal notification of key NATO allies, in particular France, West Germany and Italy, once formal agreement with the USA had been achieved. Although the original MOD intent had been for the Prime Minister to announce the decision to the House, in the event she decided that the Secretary of State for Defence should make the announcement.[44] Despite the careful choreography, things did not go according to plan. Early in the morning of 15 July, two days before the planned announcement in the UK of the exchange of letters, the UK Embassy in Washington reported that the Senate Republican leader, Senator James Baker, had informed ABC Television that the US administration had agreed to sell Trident to the UK.[45] In order to avoid the embarrassment of having Parliament find out such a crucial decision from foreign media, the timetable for informing Parliament had to be rushed forward, and Cabinet needed to be informed beforehand.

The full Cabinet was therefore informed on the morning of 15 July, in a manner reminiscent of the way Attlee had informed his Cabinet in 1948 of the effort to build a British atomic bomb, and of Churchill's announcement of the H-bomb project in 1955. John Nott (President of the Board of Trade) described the meeting:

we were simply informed by the Prime Minister that a decision had been taken in conjunction with the Americans to modernise the deterrent with the introduction of Trident. I was shocked that the Cabinet had neither been given any facts nor consulted on the issue. I protested. I said that I thought it was an unsatisfactory way of conducting the Government's business, not least because this was a matter of fundamental national importance ... The whole matter took up about ten minutes of Cabinet time.[46]

Pym made a statement to the House that afternoon and the announcement benefitted from the previous careful Cabinet Office planning, involving simultaneous briefings of the media, and a side debate among officials about whether the MOD needed to have 'more written material prepared (e.g. some sort of *Daily Mail* counterpart to the *Daily Telegraph* style of the Departmental Memorandum)'.[47] This suggests that the MOD understanding of public interest was sophisticated enough to differentiate myriad audiences, though there is no evidence that at this time MOD paid any attention to public messaging.

Pym's statement included the cost-effectiveness of Trident, UK independent operational control, the industrial impact of 70 per cent of the costs being spent in the UK, and he portrayed Trident as

an essential reaffirmation of our national commitment to security and to co-operation with our allies under the North Atlantic Treaty. The United Kingdom's continuing possession of a strategic nuclear capability remains a major element in our deterrent strategy, and a major contribution to the defence of Western Europe.[48]

The shadow Defence Secretary, William Rodgers, responded:

We have asked, first, for a full and informed debate, which has not taken place ... There are those who will say that it could be a contempt of the House for the Secretary of State to make an announcement of this sort before the Select Committee and the House have had the opportunity to discuss the matter ... We believe that the case for buying Trident has not been made, and we cannot approve it.[49]

Clearly the Opposition felt that the ability of Parliament to hold the executive to account had been circumvented in this case, but given that the Labour leadership had already been involved in the decision to procure Trident, this can be seen as little but party posturing and not a principled position.

Pym's Open Government Document 'The future United Kingdom strategic nuclear deterrent force' (OGD 80/23) was published in July 1980. This document was a significant break with previous governments' tendencies to keep official deterrence thinking out of the public eye. It contained five sections: 'The policy background' included an elementary description of NATO nuclear deterrence strategy and the unique role of the UK strategic forces in it, in particular the concept of the 'second centre of decision making' and the related concepts of credibility of a deterrence capability and resolve to use it. Section II, 'General considerations on system choice', described the philosophy behind UK targeting policy, and the related benefits of an invulnerable retaliatory strike capability (SLBMs such as Polaris and Trident). It then considered the timing and procurement of the successor to Polaris, and emphasised the risk and cost benefits of close cooperation with the USA. Section III, 'System options', considered the various options that were potentially available: air-launched cruise missiles, SLCM and various SLBM, including Polaris with a further Chevaline-style extension-of-life programme, collaboration with the French SLBM programme, Poseidon; and Trident. It starts with an objective assessment of the pros and cons of the available platforms and concludes '[f]or all these reasons, nuclear propelled ocean-going submarines remain the best launch platforms for a British missile force'.[50] The next section amounted to cost–benefit analyses of SLCM and SLBM. It describes in surprising detail the capabilities and limitations of the various existing ballistic missile systems and the existing and potentially available cruise missile options. Any one of these factors in isolation could (and would) be used to argue for and against any one system, but the document assesses the complex relationship between missile capability (range, destructive potential and post-launch vulnerability to intercept), platform vulnerability, platform numbers, cost and technological risk. Substantially, OGD 80/23 Sections I–III represented a redacted version of the Duff-Mason Report, with a consideration of the decision-making process added.

The fourth section focused on arms control, always the parallel track for UK nuclear deterrence policy. 'Strong support for practical, balanced and verifiable arms control measures remains a key element in our approach to ensuring peace and security'.[51] It indicated that the UK supported the NPT and that replacement of an existing system was incompatible neither with that treaty, nor with the US–USSR SALT I and II treaties. The final section was 'Cost':

> we assess the likely order of capital cost for a four-boat force, at today's prices, at around four-and-a-half to five billion pounds, spread over some fifteen years ... There has rightly been widespread public interest in the effect which the replacement of the Polaris force will have upon other aspects of the defence programme. Money spent on this is not money spent on other things ... Even after spending on the Trident force, the Government is still planning to spend more on conventional forces than it does now. The accommodation of large re-equipment programmes is a normal part of defence planning and budgeting ... There are no easy comparisons to be made with other defence capabilities. There would be little point, for example, in diverting the full capital sum to buying more ships, tanks or aircraft which in the long term we could not afford to run and could not hope to man.[52]

Ironically, OGD 80/23 seems to have considered many of Armstrong's proposed questions that had not been addressed by MISC 7,[53] although it did not address his first two: 'what good has it done us so far to be in the strategic deterrent league?' and; 'given the decline in our world position in other respects, will it do us enough good to stay in the league from the 1990s to justify the cost of the burden this programme represents in the meantime?' These questions are, however, biased towards the assumption that the UK retains a strategic nuclear deterrent for purposes other than deterrence.

In short, OGD 80/23 addressed the Polaris successor decision logically and objectively and should have set most arguments to rest. That it did not was manifest in the increasingly vocal and influential anti-nuclear opposition. How much of the resulting anti-nuclear activity might have been avoided had the answers in the document been published in late 1979 or earlier in 1980, and debated in the Commons before a decision (for which the Conservatives had a comfortable majority) is simply a matter for counter-factual speculation. But Armstrong's first two questions continue to bedevil UK nuclear policy decisions forty years later.

In June 1980, the Commons Defence Committee began its inquiry into the future of the UK's strategic weapons policy. Clearly it was not ideal to have the government publish its decision less than a month later, seven months before the Defence Committee reported; and there was an element of chagrin in the tone of the Report:

> Subsequently in July 1980 the Government announced their decision to purchase the Trident missile system to replace Polaris; and on 3rd March 1981 the House endorsed that decision. Since the House has voted, by 316 votes to 248, to endorse the choice of the Trident system, it is not for us to challenge the principle of that decision.[54]

There was clearly a strenuous debate within the Committee about the tenor and ultimate recommendation of the final report. The draft report prepared by the chair (Sir John Langford-Holt (Conservative)) was challenged at the final Defence Committee meeting on this topic on 20 May 1981, where an alternative draft was put forward, prepared by three of the five Labour members.

The vigour and partisan nature of the Committee's debate was exemplified by the final conclusions: the Committee's Final Report read '[w]e can see no case for the cancellation of the Trident programme by any future government'.[55] The alternative report had concluded '[w]e cannot recommend that any future Government continue the Trident programme'.[56] After the first reading of both drafts, the challenge was defeated by one vote (the eleven members voted exactly on party lines) and Langford-Holt's draft was adopted by the Committee. This was an unusually robust debate in a Parliamentary committee and although this could probably be attributed either to party-political manoeuvring or to an authentic feeling that the moral weight of the issue deserved less cursory treatment by the government, that cannot be conclusive. The previous Labour Party leadership had facilitated the replacement decision when it commissioned the Duff-Mason Report and handed this over to the incoming Conservative government. But

The Polaris replacement decision

committed unilateralist Michael Foot had been elected Labour leader in November 1980, although the party's official policy was not changed to one of unilateral disarmament until the conference of 1982.

The substantive contents of the two reports are similar, although the alternative report included a section detailing the decision-making process, which concluded:

> Parliament's role in the decision to procure a successor system to Polaris has been limited to endorsing a decision already taken ... if one concludes that nuclear deterrence is an ethically acceptable policy, it then becomes necessary to address specific strategic questions of how the threat is credibly to be made, and what political benefits are likely to accrue from the possession of a deterrent. Unfortunately, there has been much more said by British Governments about the *capability* to exercise a deterrent policy, than about precisely what policy *is*.[57]

This remains a valid criticism of every government since.

The Labour members of the Defence Committee clearly wanted the opportunity to voice conclusions and recommendations which were beyond the terms of reference of the Report, but in the event they were outvoted by the Conservative committee members. There were other, significant differences throughout the two reports in interpretation of the same evidence; and these were both symptomatic and typical of the increasingly dogmatic polarisation of views over nuclear issues as the year progressed.

John Nott replaced Pym at Defence in January 1981. To the last, Pym had sought increased engagement with Parliament and had been in negotiations with the Leader of the House for debate on nuclear policy as soon as the Commons Defence Committee had completed its Report. However, when the report was further delayed to March Pym felt that the timetable for discussion related to procuring Trident D5, rather than the C4 variant, could be delayed no further. If anything, Nott was even more enthusiastic about increasing education and understanding of nuclear deterrence matters than Pym had been and he immediately set in train a series of briefings for junior ministers and the Cabinet; although these were not popular with the Prime Minister. Nott also set about quickly mastering his nuclear brief and wrote to the Prime Minister within a month of his arrival:

> we are losing the defence/deterrence argument at present. The CND campaign is gathering strength but much more importantly, there is a growing scepticism among a much wider and thinking section of the population about the correctness of the Trident decision. If we lose the Trident argument, it will be very difficult, if not impossible, to sustain the wider defence posture of the government. I am convinced that Trident was the correct decision. But I must tell you that this is not, in my judgement, the general view of your Ministers, nor the unanimous view of the Ministry of Defence. Only the Defence Committee of our party seem free of doubts. We must win the argument in Whitehall, if we are to have any chance of convincing the outside world.[58]

The debate on nuclear policy was scheduled for 3 March 1981. Nott intended to open and close this debate because he felt that 'two Ministers can[not] be expected to have the time to master the strategic options and philosophical arguments…'[59] Mrs Thatcher did not agree, noting by hand on Nott's memo 'I really think two Ministers should be able to master this'. Nott's hesitation in trusting part of the argument to one of his subordinate ministers is entirely in keeping with what had become routine for the oversight of deterrent policy, which was kept within small groups of very senior ministers and not even divulged to the full Cabinet until decisions had been made.

In the run-up to the debate Nott planned to schedule a number of in-depth interviews with television and newspapers; 'we have had many useful requests to enter a serious debate on the moral and strategic issues'.[60] This appears not to have happened until after the Parliamentary debate, with Nott announcing in an interview with *The Times* that he had 'decided to wage a public relations battle to wean "innocent well-meaning people" away from the ideas of the Campaign for Nuclear Disarmament …'[61]

In early 1981, despite coverage of the ongoing Commons Defence Committee hearings on the future of the UK strategic weapons policy, there was no apparent increase in government participation in the public discourse. In the meantime, opposition was growing and CND was attracting more members, increasing membership tenfold between 1979 and the end of 1982.

The Polaris replacement decision

Nott opened the debate as planned, and took the non-technical issues of nuclear policy head-on:

> In ethical terms, the issues surrounding nuclear weapons are difficult, if not agonising. But in the debate between those who judge it better to keep those terrible weapons so as to use them as a shield for peace, and those who judge it better to discard them in order to maintain peace by some new, untried and, I would suggest, historically improbable route, the arguments are, I would suggest, on the side of deterrence. To engage the emotions – as the promoters of CND know very well – is an easy task. The showing of the film *The War Game* in a village hall in the evening in the presence of young families has a predictable outcome. To argue the choices before us so as to engage the intellect is a much harder task.[62]

Nott considered again the key decision factors outlined in the Duff-Mason Report, and already covered in the previous debates; including in particular the relevance of an independent centre of decision-making, the minimum deterrent and the irrelevance of parity with the Soviet Union. He concluded his introduction:

> All of us fear and abhor the idea of war, and, above all, of nuclear war. All of us have a common aim – to prevent it, but it must be deterred away or negotiated away; it cannot simply be wished away. Britain has a distinctive role in deterrence – one that our allies acknowledge and welcome. It would be dangerous folly, in the world as it is, now to abandon that role. Much the best long-term way to sustain it at the strategic level is to build a new force around the Trident missile. That is, in absolute terms, not a cheap course, but the consequences of shirking it may one day prove unimaginably expensive.[63]

Shadow defence spokesman Brynmor John also had laudable intentions, to face the key issues and eschew party politics, but as the debate unfolded it became clear that there would be a bi-partisan division between Conservative and Labour on the nuclear deterrent for the first time since the deployment of Polaris. This was articulated in unequivocal terms by Labour MP Peter Snape:

> The Labour Party is against Trident. It ill-behoves the Conservative Party, which has cut, cut and cut again in relation to the social structure of this country, to spend a minimum – and Conservative Members know that it is a minimum – of £5 billion on the Trident

project. Those of us who are concerned for not only the future of Britain but the future of our children are determined to ensure that there will be no Trident project in the future. My hon. Friends and I are the vanguard of a movement which has a great depth of feeling throughout the country.[64]

The Commons endorsed the government's decision to maintain a strategic nuclear deterrent and the choice of the Trident missile system as the successor to the Polaris force. The technical factors changed again within the year when the USA decided to proceed with the early replacement of Trident C4 with the D5 variant. The UK was faced with almost the same decision once again – change the plan and provide a system which would remain compatible with that deployed by the USA throughout projected design life, or take the (slightly) cheaper option and face increasing obsolescence and UK-specific upkeep costs.

Anti-nuclear opposition focused on both the decision to replace Polaris and the planned deployment of cruise missiles in 1983. CND membership rose from 4,000 in 1979 to 20,000 in 1981 and peaked at 100,000 in 1984. Turnout at the annual demonstrations in the early 1980s regularly topped a quarter of a million. Peace camps were established, most notably at the proposed missile bases at Greenham Common and Molesworth (although CND's involvement at Greenham was tangential). A wide variety of non-violent direct action took place, usually at military bases and nuclear weapons manufacturing/storage sites. Whilst some was organised by CND at a national level, the norm was for local initiatives, with national CND coordinating rather than directing.[65] The National Executive Committee of the Labour Party, which had voted for unilateral nuclear disarmament at its 1980 party conference, was one of the sponsors of the 1980 demonstration against cruise missiles.[66]

Media coverage of the large march from Aldermaston to Trafalgar Square on 26 October 1980 tended to indicate that the media would find themselves unable to avoid taking a side in the rapidly polarising debate. The *Daily Express* reported the rally itself objectively:

More than 60,000 people turned up yesterday for the biggest nuclear disarmament demonstration since the 'Ban the Bomb' demo's *[sic]* of nearly 20 years ago. Their message was – no Cruise missiles! No Trident submarines! And a massive cut in arms. It was part of the re-birth of the peace movement and the support overwhelmed organisers.[67]

The same article also covered Michael Foot's statement to ITV's *Weekend World*:

> Mr Foot said that if he became Prime Minister he would send America's Cruise and Pershing missiles home – and get rid of Britain's independent nuclear weapons ... Labour defence spokesman William Rodgers came out against Mr Foot's plans. He said: 'Michael Foot's remarks were a plain statement of unilateralism. This is not the majority view of the parliamentary Labour Party. I do not believe it is the majority view of Labour voters. The people of this country believe that Britain should be properly defended.'

The *Daily Express* also carried an editorial considering Foot's commitment should he become leader of the Labour party and potentially Prime Minister: '[h]e is prepared to do away with Britain's nuclear weapons unilaterally. So the man whom we are being invited to consider as our next Prime Minister is a man who would deprive Britain of the decisive weapon of the modern age.'[68] The *Times* editorial commented:

> the massive demonstration of support for the Campaign for Nuclear Disarmament on Sunday afternoon was an event of considerable political significance because it represented the revival of a movement whose activities had such an impact on British public affairs at the beginning of the 1960s. Why has this revival occurred now, and for how long can one expect it to last?[69]

The investigative documentary programme *TV Eye* predicted that in the event of a war, the 'first Nuclear missiles would fall on Britain'.[70] Interviewing Francis Pym – a relatively rare public intervention by government – reporter Bob Southgate put together a map of American bases in Britain and asserted '[t]he American presence here is so large and so important that the Russians could feel forced to attack these bases first'. Pym countered that 'both a British and American nuclear arsenal are vital to the defence of the West and to stop nuclear war becoming a reality'.[71]

The Times also sketched the reduction in interest in CND after the Cuban missile crisis and suggested:

> that has changed with the dispute over the Trident and Cruise missiles. The argument has been joined once again, with a public many of whom are totally uninfluenced by the previous debate and in circumstances that are different in a number of respects. One is that the international scene seems more forbidding ... Recognition of this threat is seen in the new preoccupation with civil defence, which is both an acknowledgement of the danger and to many people an inadequate safeguard against it.[72]

As well as the 'traditional' mass demonstrations, anti-nuclear protest took novel forms. The Welsh group Women for Life on Earth marched from Cardiff to RAF Greenham Common where the first detachment of ninety-six GLCM would be based, arriving on 5 September 1981. They intended to challenge the decision by staging a debate with the base commander. On arrival they delivered him a letter which among other things stated '[w]e fear for the future of all our children and for the future of the living world which is the basis of all life'.[73] They were denied the debate, but took up residence in several makeshift camps on common ground adjoining the base, and remained there as a highly visible focus for the anti-nuclear protest movement until 2000, long after the GLCM had been withdrawn in 1991. Similar camps were established at RAF Molesworth, RAF Alconbury and at Faslane, the Clyde base from which the Polaris (and subsequently Trident) submarines operated.

These protests, committed to disrupting the daily working lives of the nuclear establishments, drew on non-violent tactics for anti-nuclear protests that had previously been avoided. Nuclear convoys leaving the bases were blockaded, tracked to their practice areas and their training and exercises were disrupted. Protesters would challenge security, cut wire fences and, if they could gain access, damage equipment within the perimeters of these facilities. This move to non-violent direct action meant that protesters were arrested, taken to court and occasionally imprisoned. Between March 1981 and 1984, over 5,000 were arrested across the country, with 1,000 arrested at Greenham alone.[74]

The Polaris replacement decision

There was little that could have made the public presentation of British nuclear policy more demanding in 1980, but the Home Office managed to find it. After the Strath Report in the 1950s, and continued decline throughout the 1960s, civil defence had been put into a 'care and maintenance' state in 1968. However, shortly after the election in 1979 the Home Office commissioned a report on 'civil preparedness' for the Defence Overseas Policy Committee (DOPC), a Cabinet sub-committee. This secret report highlighted the parlous state of civil defence and concluded that

> [t]here is widespread ignorance among the public, the media and officials of the threat and about protective measures. For many years the perception has been of immediate general nuclear war. Until recently this has led to apathy in the face of the appalling consequence and of the perceived inadequacy of steps taken to alleviate them ... A more open approach by government now would counter criticism of undue secrecy and a more informed public might change its attitude. It has been announced to Parliament that the pamphlet 'Protect and Survive' will be updated and placed on sale at the time the outcome of this review is announced.[75]

Cabinet endorsed the report's recommendations and advised the Home Secretary to '... aim at an undramatic statement, in order not to arouse expectations which it would be impossible to fulfil'.[76] Prior to the announcement of the report, the Chief Whip was worried: 'I have not seen a copy of the paper which we shall be publishing, but I get the impression that it may cause a good deal of trouble by its inadequacy'.[77] In June, the DOPC agreed that the Home Secretary should announce the report and its recommendations, and that the 'financial problem should be considered further in the forthcoming Cabinet discussion of the 1980 Public Expenditure Survey with the aim of implementing all of the Priority I and as many as possible of the Priority II measures'.[78] This left Whitelaw in a nearly impossible position because the costs of most measures to protect the population from the consequences of a nuclear attack were prohibitively high, and the measures were probably ineffective, which was why they had been abandoned thirty years before and the focus had been shifted to deterring rather than fighting a nuclear war. Whitelaw's task now was to present the cheap options as part of a package designed to educate an increasingly sceptical public.

'Protect and survive; civil defence manual of basic training' had been issued to the Civil Defence organisation in 1950.[79] It incorporated all of the previously published pamphlets on basic chemical warfare, firefighting, first aid, rescue, protection from high explosive missiles and atomic warfare. It was re-issued to the Civil Defence organisation in various forms until 1968; Civil Defence handbook No. 10 'Advising the householder on protection against nuclear attack' was issued in 1963.[80] It included advice on protective measures such as building a fall-out room under the stairs or the kitchen table. The decision to update and reissue Handbook No. 10 as a public document in 1980 was a disaster for the public presentation of government nuclear deterrence policy.

Based on the experience of the 1960s, the decision to publish this document displayed a degree of ineptitude verging on the imbecilic. 'The government's civil defence pamphlet, Protect and Survive – perhaps the greatest own goal of the 1980s – was a gift to the peace movement.'[81] '[It] sought to persuade people that they could protect themselves against bombs with the power of millions of tons of dynamite by crouching under a table'.[82] In response, CND produced a pamphlet entitled 'Civil defence: the cruellest confidence trick', which described 'Protect and survive' as 'a mass confidence trick, a public fraud of the most heartless kind because it deals in human lives'.[83] CND seized on the title of the civil defence booklet and corrupted it into the campaign slogan 'Protest and survive', which became ubiquitous, for example as the title of a photograph of a skeleton reading 'Protect and survive' which was displayed at the Tate Gallery.[84]

In 1980, the National Council for Civil Defence started planning for Exercise Hard Rock, the first Civil Defence exercise to consider a nuclear attack for nearly thirty years. However, after twenty-four of the fifty-two councils refused to participate, Whitelaw was forced to cancel it. According to one press report, 'just 55 men and women are expected to handle all aspects of peacetime emergency and wartime emergency planning for more than 18,500,000 people ... In Tyneside, one person would look after 1,100,000 people.'[85] This *Times* article presented the government attempts to improve the civil defence situation favourably, and actually condemned CND for its critical attitude, but in doing so it highlighted the position baldly; either significant investment would be required, or the civil defence project was, as CND General Secretary Bruce Kent put it, ludicrous.[86] By its very objectivity, the *Times* article enhanced the CND argument.

The public reaction to civil defence was pretty much that which had been predicted in 1955 in the Strath Report, hence the extreme secrecy which still surrounded that document, and in the persistent concerns over *The War Game* in 1965. Fiction was also catching up; in the USA, a TV movie, *The Day After*, presented the aftermath of a nuclear war in the contemporary genre of a disaster movie, attracting the (then) largest ever TV audience (100 million) when it was first screened in 1983. Prior to airing, there was intense debate in the USA whether it should be screened, with the 'Educators for Social Responsibility and others worried that the program might do children more harm than good'.[87] Immediately after the film was shown, *Viewpoint* hosted Henry Kissinger, Robert McNamara, Brent Scowcroft, Carl Sagan and others for a ninety-minute panel discussion with a live audience:

> Contrary to [host] Koppel's assumption that the ... 'simple-minded' film would do little more than encourage the nation to 'make policies by scaring ourselves to death,' the audience members demonstrated themselves to be calm, rational and well versed in international affairs ... They were clearly not the easily frightened or brainwashed masses conservatives feared or the eager converts hoped for by disarmament activists.[88]

In 1984, in the UK, the BBC produced *Threads*, depicting the destruction of Sheffield in a nuclear war. It was of the same genre as *The Day After*, although it depicted far more of the run-up than the war itself, but in doing so it illustrated the utter irrelevance to the film's characters of the issue over which the war is fought. The film contained several aspects paying 'homage' to Watkins's *The War Game*; the 'authorities' come out very badly – shooting looters out of hand and holding summary trials – while the 'legitimate' local government is buried alive in a bunker and completely ineffectual. The film follows the daughter of the main characters, born five months after the bomb, to the point at which she gives birth to a still-born baby at age 14. By this stage, British society has regressed to a neo-feudal state and her vocabulary is limited and sounds like medieval English. Films such as these graphically portrayed the 'breakdown' of society which had so worried Strath. The concepts of protection portrayed in 'Protect and survive' were most ruthlessly parodied in Raymond Briggs's book *When the Wind Blows* and the subsequent animated film of the same name.

Contemporary 'teen' fiction in the early 1980s also exploited the aftermath of nuclear war for plot lines; *Children of the dust* recounts the stories of three individuals in successive generations of survivors.[89] There is no depiction of the nuclear war, or mention of its cause, simply the creeping inevitability of death caused by 'dust'. The government-imposed society fails because it cannot adapt to the new reality and simply tries to reimpose the previous technocracy and its long-defunct political legitimacy on the survivors. It is as bleak as any of the adult fiction initially, but where the adult fiction tends to leave the reader either witnessing the death of all of the characters, and perforce mankind, this offers a more positive outlook, albeit in a pseudo-science-fiction vignette with mankind evolving into a more peaceful post-holocaust humanity. Similarly, *Brother in the land* focuses on the development of society immediately after a nuclear attack, suggesting it reverts to 'pre-Neanderthal';[90] once again suggesting Strath's 'breakdown'. These children's books painted a very bleak appraisal of humanity in a state of nature, and implicitly challenged the assumption that defence of a particular model of society could justify a nuclear war.

As Grace Wyndham Goldie had said in 1965, 'so long as there is no security risk and the facts are authentic, the people should be trusted with the truth ...'[91] The government's continued reluctance to provide the difficult facts to the public could do little but foster a sense of intrigue and fuel the CND narrative that the authorities had something to hide. Fear caused by increasing East–West tension in the early 1980s, exacerbated by these portrayals of nuclear war and the breakdown of society, was almost certainly a contributory factor in the increasing popularity of CND. That said, opinion polls did not suggest that the overall public view of nuclear disarmament was significantly affected. Philip Sabin's detailed analysis of the poll results suggested that the perception that a war would be either a full-scale conventional battle in Europe or a nuclear holocaust did affect fears associated with nuclear weapons, though there was a majority view against unilateral disarmament on the basis that possession deterred war in the first place. Of note, Sabin's analysis concluded with a speculation: '[d]id growing fear of war spark off the recent security debate, or has the recent war scare itself been a by-product of increased discussion, acrimony and alarmism about the defence issue?'[92] This further suggests that the lack of definitive guidance and intervention by government allowed any public discourse to be dominated by the anti-nuclear narrative.

Nott had noted three years before that the opposition to nuclear deterrence was essentially emotive, and that the government problem was how to counter it with rational points. He appears to have missed the most important point, though – the opposition set the parameters of the debate, and anchored it in these highly emotive depictions of the aftermath of nuclear war. Unchallenged, these perceptions contributed to the disarmament narrative – the government could have simply agreed that all war (and nuclear war in particular) was a dreadful prospect and then described the various options to avoid it, of which nuclear deterrence seemed the most viable. For whatever reason, this line was not successfully pursued.

In April 1963, Pope John XXIII had published the Papal encyclical *Pacem in Terris*, which stated:

> Justice, right reason and consideration for human dignity and life urgently demand that the arms race should cease; that the stockpiles which exist in various countries should be reduced equally and simultaneously by the parties concerned; that nuclear weapons should be banned; and finally that all come to an agreement on a fitting programme of disarmament, employing mutual and effective controls.[93]

This directive to the world's Catholics had little impact on national nuclear deterrence policies at the time, but it set a precedent that had significant effect decades later.

With concern mounting over the stand-off in Europe over long-range tactical nuclear forces (LRTNF) in the late 1970s and early 1980s, pastoral concerns overcame political reticence and a number of churches chose to voice opinions on nuclear deterrence. The prompt for this was the intervention of Pope John Paul II at the UN Special Session on Disarmament in June 1982: '[i]n current conditions "deterrence" based on balance, certainly not as an end in itself but as a step on the way toward a progressive disarmament, may still be judged morally acceptable'.[94]

Probably the most famous development from this is the Pastoral Letter of the American Catholic Bishops of 1983: '[a]s Catholic bishops we write this letter as an exercise of our teaching ministry ... We do not perceive any situation in which the deliberate initiation of nuclear war, on however restricted a scale, can be morally justified.' As well as instigating a cottage industry in debate whether or not the Bishops' views were a valid representation of the Catholic Church position, this pastoral letter presaged a number of similar exercises by other churches. In the UK, the Scottish churches were vigorously opposed to the Polaris replacement and LRTNF decisions:

> For 30 years the Church of Scotland has consistently condemned the existence and threat of nuclear weapons as sinful and an offence to God's created order.[95]
>
> If it is immoral to use these weapons it is also immoral to threaten their use. Some argue that the threat can be justified as the lesser of two evils. The crux of the problem is whether in any foreseeable circumstance a policy of self defence based on the use or even the threat of use of these weapons of terrible destructiveness can ever be morally justified.[96]

The English churches took a more measured approach. The Catholic Bishops invited the Cabinet Office to brief their 'In Service Course' on the matter in January 1983: the Cabinet Office official who briefed them reported '[o]n the question of principle, the Bishops were predictably wrestling with the morality of conditional intention: i.e. could it ever be right to have the intention to commit a morally monstrous act, however justifiable and desirable the objective?' He suggested that the key debate among the Bishops was ethical; the extent to which deterrence was stable and therefore the conditional intent to use nuclear weapons could be considered to be at a very small risk of realisation. He also pointed out:

> They seem to have been impressed by a presentation which they had received the previous day from Dr Paul Rogers of the Bradford University School of Peace Studies, who had apparently argued in

favour of Britain conditionally offering to abandon Trident (and Polaris) in return for some matching move from the Soviet Union.[97]

As described above, Michael Quinlan engaged with some of the protagonists of this debate in a personal manner (but with tacit official sanction):

> In his 1981 pamphlet 'The Morality of Nuclear Weapons', Ruston ... argued that: 'There is no way in which the present possession of nuclear weapons, intended for whatever purpose, can be justified in Catholic morality'... This was refuted by Quinlan, who argued that surrendering to an atheistic totalitarian regime would be an immoral act.[98]

Quinlan remained active in a private capacity; as a deeply committed Roman Catholic himself, he thought deeply about the ethics of nuclear weapons and sought constantly to overcome misperceptions and ignorance on the subject in his widespread correspondence, until his death in 2009.

Given the febrile atmosphere, the Church of England issued a moderate pamphlet 'The church and the bomb' and passed a relatively constrained motion at the General Synod of February 1983 which recognised:

> it is not the task of the church to determine defence strategy but rather to give a moral lead to the nation;
>
> (i) Affirms that it is the duty of Her Majesty's Government and her Allies to maintain adequate forces to guard against nuclear blackmail and to deter nuclear and non-nuclear aggressors ...
> (ii) Judges that even a small-scale first use of nuclear weapons could never be morally justified ...
> (iii) Bearing in mind that many in Europe live in fear of nuclear catastrophe and that nuclear parity is not essential to deterrence, calls on her Majesty's Government to take immediate steps ... to reduce progressively NATO's dependence on nuclear weapons and to decrease nuclear arsenals throughout the world.[99]

Government engagement with the churches on nuclear matters in England remained amicable throughout the 1980s, although individual churchmen of personal principle were intimately involved in the Campaign for Nuclear Disarmament and other protests. The Scottish churches were much less cooperative and have retained a much more radical anti-nuclear stance ever since. One subsequent church analyst concluded:

> [The bishops] want to be taken seriously and not dismissed as extremists or unrealistic. But the criteria of Just War teaching are not unrealistic or irrational, and many men and women – not just Catholics – would rejoice to hear Catholic doctrine affirmed as loudly and unambiguously concerning war as it has been in other discourses.[100]

Ultimately, as the Anglican synod agreed, the role of the church is not to determine defence strategy but rather to give a moral lead to the nation. The British church critiques that were presented to the public tended to be deontological rather than consequentialist, therefore tended to contribute to polarisation of any subsequent debate and did not contribute to consideration of alternatives to deterrence. They also tended to conflate deterrence and war: 'Pope John XIII stated that in an age of atomic power "it is irrational to think that war is a proper way to obtain justice for violated rights". The Vatican Council emphasises "the unique hazards" of modern war arising from the weapons now available.'[101] This is not a critique of deterrence, but of war, the avoidance of which is the key function of deterrence. This is another example of a key argument being corrupted by a superficial understanding of the full scope of the factors being considered; an argument that could have been substantially answered by more informed official participation. The relevance of this discussion for nuclear deterrence in the 21st century will be explored in chapter six.

During 1981, the MOD progressed negotiations and plans for the procurement and construction of the successor system to Polaris. MISC 7 remained occupied (in secret) with the decision to procure Trident II D5 instead of the C4 system, offering cheaper through-life costs due to sustained commonality with the US Navy throughout the system's projected life. In November, MISC 7 was ready to decide in favour of D5, and the Cabinet was informed on 21 January 1982. It was invited to decide on D5 on 11 March, although in order to enable American Congressional briefings before the announcement on the same day, Cabinet made the decision on 3 March and Nott announced the decision to the Commons on 11 March. Trident II D5 was far more accurate than the C4, and offered greater range and more warheads per missile. There was therefore potentially a presentation issue – D5 offered far more firepower than the British minimum deterrent required, and it could be portrayed as a 'warfighting', rather than deterrence, system.

This point was made in the Commons reaction: '[d]oes the Minister agree that it is not simply a replacement programme but that in terms of quality and quantity of warheads D5 represents almost a quantum leap forward? Will the Minister tell the House why we require a hard kill capability?' Nott responded: '[a]s for the hard kill capability, it is certainly true that D5 is a more accurate missile than C4 and much more accurate than Polaris ... But that is not why we want it. We have chosen it because of commonality with the United States.'[102] This distinction added a further focus to the polarising effect of the Trident debate in public.

The Falklands War diverted much of the media and protest attention away from the nuclear debate, and over the summer MISC 7 continued to explore means of reducing costs such as cancelling the expansion of Coulport (part of the submarine base on the Clyde) to process the Trident missiles and negotiating use of American facilities instead. This would save £500 million but cost 2,000 jobs in Scotland over the following fifteen years. The Cabinet Secretary also noted that this move might be perceived as a loss of independence in the deterrent.[103]

In his routine 'Media relations stocktake' for Mrs Thatcher in August 1982, her press secretary Bernard Ingham did not mention nuclear deterrence as an issue at all: 'it is possible to identify the main ... issues which will preoccupy us over the next 12 months: the Franks report [into the events leading to the Falklands War] ... unemployment, pay, inflation, trades union reform ... public expenditure ... crime and punishment ... membership of the EC'.[104] Ingham's subsequent preparations for the September 1982 meeting of the Lord President's Liaison Committee, which was developed 'to give guidance to MPs and Ministers on the interpretation of Government policy and to take such action as in their opinion is necessary to sustain public confidence in Government', included presentation on economic issues, law and order, housing and social security policy, with potential for a paper on nationalised industry.[105] He seems to have completely misread the priorities; in the event, the key agenda item was the identification of a nuclear policy presentation strategy:

> the need for concerted effort to secure public acceptance of the Government's Trident decision. Opponents were able to range widely in their criticism and to quote in their support sensible people who were concerned with e.g. conventional capabilities or industrial implications. The CND campaign was likely to constitute a serious and continuing problem. A great deal of valuable material had already been issued by the Government to explain its decisions in this area and Conservative Research Department should draw on what was already available in the preparation of a consolidated paper on the presentation of defence and nuclear policies ...[106]

This meeting appears to have set in train a number of changes in the way that the Conservative government dealt with nuclear policy in the public domain. This marks a shift in strategy; until this meeting, UK nuclear policy was presented in public only if it had to be, in a grudging, haphazard and often counter-productive manner; now it was perceived as a serious domestic-political issue and, on party political grounds, the public messaging needed to be far more coherent. At the October meeting, the Conservative Research Department paper on the presentation of the government's nuclear defence policy was the only substantive agenda item. It reported that:

> b) Opinion polls suggested that a majority of the electorate were broadly in support of the retention of a proper military capability; but only a minority supported Trident even among the Government's own supporters ...
> c) The unilateralist lobby had grown very strong and very influential. It seemed to be making an impressive impact not only among the general public but also in the universities and at local authority level ... It was essential to counter such campaigns effectively at the local level.
> d) In presenting its decision on Trident the Government had naturally set the issue in the context of the threat from the Soviet Union ... The Government's defence policies needed to be

The Polaris replacement decision 131

presented in the round and set in a wider context than NATO alone.

e) The wider work done by the Ministry of Defence to counter the unilateralist lobby was noted, as also the activities of outside associations which had been set up for the purpose, or were prepared to help. But it was not clear that the Ministry of Defence had achieved a sufficient impact. Some of its published material lacked appeal. The Government's message was not getting over to the general public or to opinion formers as strongly as it should. There appeared to be inadequate information officer effort devoted to this task ...

f) CND publications were simple, emotive and effective; and they were pushed hard by their supporters at all levels. To redress the balance there was much to be said for harnessing the energy, imagination and enthusiasm of the Party organization ... They could properly undertake the sort of presentational campaign which it would be improper for a Whitehall department to organise. But they lacked financial resources.[107]

This, finally, was a recognition of Nott's lament to the Commons eighteen months before: '[t]he showing of the film *The War Game* in a village hall in the evening in the presence of young families has a predictable outcome. To argue the choices before us so as to engage the intellect is a much harder task.'[108] The Liaison Committee recognised that this area of presentation stood in need of urgent review, both in substance and method. 'Renewed efforts had to be made to present the message in appealing and effective terms and to get it over strongly to opinion formers ... The Ministry of Defence needed to reassess the staff and financial resources allocated to this important job.'[109] This Conservative Party committee recommended a full Cabinet discussion and the Minister of State for Defence was invited to report back to the Secretary of State.

Defence Secretariat 17 was a small, long-standing section within the MOD which was tasked to advise the Secretary of State for Defence on advice on nuclear policy, arms control and disarmament. In February 1983, Heseltine (who had taken over at the MOD from Nott) established a parallel eight-man unit, Defence Secretariat 19 (DS19), to advise him on how best to explain to the public the facts about the government's policy on deterrence and multilateral disarmament; its work was to be confined to departmental information work and specifically 'it will be assisting in the preparation of speeches and articles, advising on leaflets and publications and advising on press aspects of the nuclear debate'.[110] In the Lords, debate arose on whether DS19 was involved in the presentation of *government* policy or, because the Opposition was publicly opposed to that policy, the unit was in fact presenting *party* policy. Lord Jenkins (Labour) inquired:

> Will the noble Lord accept that sometimes it is rather hard to draw the line between legitimate Government explanation of their policies and party political propaganda? ... Will he further agree that some of the films which are being issued under the aegis of this department [DS19] are highly contentious, and that to this side of the House they feel like party political propaganda?[111]

Lord Belstead neither accepted nor agreed.

By April, this more proactive government campaign had significantly shifted the public discourse and Heseltine reported to the Cabinet that 'the Campaign for Nuclear Disarmament (CND) had been successfully thrown on to the defensive by the action taken to identify the left wing affiliations of so many of its leading members'.[112] *The authorised history of MI5* asserts that in the mid-1970s eight of the fifteen seats on the CND National Executive Council were occupied by members of the Communist Party and '[i]n March 1983 the Service [MI5] provided the MOD with open-source material on the political affiliation of seven leading members of CND'.[113]

Anti-nuclear campaigners maintained a steady pressure; Joan Ruddock (CND chair) recalled '1983 was general election year and attacks on CND escalated dramatically. We repeatedly asked to debate with ministers but to no avail.'[114] In 1985, an MI5 case officer approached Channel Four's flagship investigative programme *20/20 Vision* alleging that, in violation of their own rules, MI5 had been conducting surveillance on CND leadership and senior leaders of the miners' unions.[115] The programme was scheduled to be aired in February 1985 but was cancelled on the advice of the Independent Broadcasting Authority, on the basis that it 'had been advised by counsel that it would be committing a criminal act under the Official Secrets Act if the programme were shown and as it was a statutory body responsible to Parliament, it should not deliberately break the law'.[116] Amongst other claims, the *Guardian* article just quoted asserted that MI5 had passed to DS19 'non-classified information about the left-wing affiliations of senior members of CND, which, it is claimed, Mr Heseltine used in political speeches about the peace movement'. The programme was shown to MPs, and the Home Affairs select committee demanded an inquiry into the activities of 'Special Branch' and MI5.

This is consistent with the *Authorised history of MI5* which records surveillance of the Greenham Common peace camps and various members of CND leadership (including Kent and Ruddock) between 1982 and 1985. This history also records – which of course would have been unknown to *20/20 Vision* and CND – that 'KGB directives passed by Oleg Gordievsky to [the UK Secret Intelligence Service] after he arrived at the London residency in the summer of 1982 demonstrated that Moscow regarded the anti-nuclear movement ... as "our natural allies" and believed it could exercise considerable influence over it'. The history further describes how Gordievsky's allegations were moderated by senior MI5 officials in briefings to the Prime Minister 'because of fears that the Prime Minister would take too literally exaggerated KGB claims of its ability to influence the movement'.[117] Ruddock noted:

> It was spine-chilling stuff. Of course we always suspected, but to see it in print was just sickening. An MI5 whistleblower, Cathy Massiter ... detailed MI5 phone taps, agent infiltration and the use of MI5 material for party political purposes.[118]

This raises the questions, important in a liberal democracy, how far is it a responsible use of the security services to exercise surveillance of those who break the law to carry out such activities, (a) however morally motivated the watched may be and (b) whether or not the watchers operate under 'appropriate' oversight and accountability, however defined? Massiter's key allegations were that accountability had been lacking, not that the surveillance *per se* was wrong. A subsequent government inquiry found no such failure:

> The Prime Minister was personally 'very concerned' by Massiter's appearance on television and asked Sir John Jones for an 'absolute

assurance that there had been no unauthorised interception of subversives'. The DG replied that there had been none since 1972 when it was within his knowledge.[119]

Lord Bridges was tasked with an inquiry, the output of which was labelled a 'complete whitewash' by CND.[120] Although some concern was expressed by the Bridges Inquiry about increasing politicisation of the security services,[121] the narrative above does tend to suggest that MI5 may have played down intelligence evidence because of potential political interpretation: always an issue when political interests are exposed to raw intelligence rather than analysis.

The Labour Party, which suffered a crushing defeat in 1983, had adopted a unilateralist stance during this election, its manifesto being described as the longest political suicide note in history.[122] Even Kent, the General Secretary of CND, was quoted saying '[w]e were badly let down by the spokespeople of the Labour Party'.[123]

Historical evidence – conclusions

The chapters so far have considered the historical evidence for government engagement in public discourse on nuclear deterrence policy. The complex relations between public perception and presentation, nuclear policy, nuclear strategy, technical factors such as cost and performance and *Realpolitik* are played out in the most real of circumstances, for the very highest stakes. The relations between these issues have second and third order implications which could be significant but which are seldom addressed when engaging in public policy. They do, however, tend to shape that engagement.

Capability of nuclear weapons is a relative measure; an atomic bomb is hideously powerful, but ineffective in comparison to a hydrogen bomb. But when combined with missile technology, and then ABM technology, the arms race associated with maintaining the credibility of the nuclear deterrent during the Cold War meant significant resources were devoted to research and development at the very cutting edge of science and technology. This drove significant costs into the maintenance of a deterrent which would remain credible (capable in relation to the threat environment) in the eyes of an adversary whose own technology was advancing. The notion of credibility will be explored more fully in chapter seven.

Technical factors were always a consideration in themselves, but they also influenced strategy and policy decisions. Assuming cost is a technical factor driven by technological demands, the need to provide a credible capability to assure deterrence drove the decisions to purchase Skybolt, Polaris and finally Trident. The need to purchase American systems was driven by the inability of the UK to sustain the research and development base because of the crippling cost. The Chevaline project was the epitome of this, and the last example of a British sovereign nuclear weapon system development project, and even Chevaline was extensively supported with American facilities and expertise.

Military factors also played a part as technical imperatives. Successive governments treated NATO and the sovereign strategic nuclear deterrent capabilities as logically distinct, although alliance commitments were used by Wilson to support the case for Polaris in 1964. This mindset hindered the Thatcher government's response to the simultaneous protests over NATO intermediate-range nuclear forces and the replacement of Polaris.

All governments handled nuclear deterrence policy decisions in very limited groups, with membership restricted to only the most senior ministers or officials; this was due in part to an inherited, obsessive secrecy, and in part to a feeling that only senior ministers could be trusted to consider and present the sensitive national security arguments in their ethical context; scruples, as Churchill put it.

Some of these scruples have significantly influenced the nature of government engagement in public discourse on nuclear deterrence. They appear to have their roots in the First World War, and to have been honed in the desire to appear to retain the moral high ground during the Second World War, to the extent that the UK's real bombing policy was not revealed to the public. A similar aversion to admitting the threat to non-combatants appears to have inhibited discussion of the nuclear deterrent since. Given the influence of this ethical reticence on the development of public engagement, chapter six will consider specifically the relationship between government nuclear policy and current ethics of war.

In the early 1980s, the Conservative government was concerned for how its message about nuclear deterrence was getting across to the public in the very febrile environment fuelled by the stand-off between the USSR and NATO over intermediate-range nuclear forces in Europe and the simultaneous decision to replace Polaris with Trident, and the resurgence of CND. Aggressive government campaigning targeting both the inherent arguments of nuclear deterrence in the Cold War, and the political sympathies of the leaders of the anti-nuclear opposition, was successful in averting a major shift in public opinion away from the government support for an independent nuclear deterrent. Despite some false starts, in particular the spectacular own goal over civil defence, the Conservative government started its second term in office with a clear mandate to procure Trident and to deploy the NATO ground-launched cruise missiles.[124]

The successful Intermediate-range Nuclear Forces (INF) Treaty of 1987 was due in no small part to the determination of President Reagan and Premier Gorbachev, but Reagan's position was supported in no small measure by this alliance-wide *Realpolitik* measure. The campaigning of European anti-nuclear demonstrators did not contribute to this success; it was the eventual deployment of GLCM and Pershing SRBM that forced the Soviets to the INF negotiations; ironically, the determination that the British, German and Dutch governments had shown in facing down their domestic opposition to the deployments must have indicated to Moscow the resolve within the European host states of the alliance to achieve either parity or removal. The timidity the governments of those nations show today in addressing nuclear deterrence matters in public may well have the opposite effect. In short, a government that struggles to articulate its nuclear deterrence policy to its own populace is unlikely to appear overly credible to a potential adversary. The Thatcher government learned this, over three bruising years.

Notes

1. Conservative Party, general election manifesto (London, 1979), ch. 6.
2. Cabinet Office. Memorandum, Hunt (Cabinet Secretary) to Prime Minister, A09454, 'The future of the deterrent', 4 May 1979. TNA 19/14.
3. 'Weapons of mass destruction; SALT II', Federation of American Scientists. Available: https://fas.org/nuke/control/salt2/index.html [accessed 13 January 2021].
4. MOD, memorandum, deputy under-secretary (policy) (DUS(P)), 436/79 'Long range theatre nuclear forces', 6 July 1979.
5. Ibid., para. 2.
6. Cabinet Office. 'The future of the deterrent'.
7. Cabinet Office. Memorandum, Hunt (Cabinet Secretary) to Prime Minister, A09588, 'The future of the deterrent', 18 May 1979. TNA PREM 19/14.
8. For detailed consideration of the 'Moscow criterion' and its impact on nuclear deterrence policy in the UK, see Stoddart, 'Maintaining the "Moscow criterion"'; Baylis and Stoddart, 'The British nuclear experience: the role of ideas and beliefs (part one)', *Diplomacy & statecraft*, 23 (2012), 331–46; and '(part two)', *Diplomacy & statecraft*, 23 (2012), 493–516.
9. Cabinet Office. 'The future of the deterrent', 18 May 1979.
10. Cabinet Office. Memorandum, Armstrong (Cabinet Secretary) to Prime Minister, A0500, 'Future of the strategic deterrent', 29 October 1979. TNA PREM 19/14.
11. Cabinet Office. Memorandum, Armstrong (Cabinet Secretary) to Prime Minister, A0547, 'Future of the strategic deterrent', 2 November 1979. TNA PREM 19/14.
12. During the *Spycatcher* trial in 1986 Sir Robert Armstrong was credited with coining the phrase – in his original, it does not mean telling lies, simply not revealing the whole truth …
13. Cabinet Office. 'Future of the strategic deterrent', 2 November 1979.
14. Cabinet Office. 'Future of the strategic deterrent', 29 October 1979.
15. *The Guardian*, 'UK ready to buy Trident missiles' (1 November 1979).
16. Cabinet Office. Memorandum, Armstrong (Cabinet Secretary) to Prime Minister, A0548, 2 November 1979. TNA PREM 19/14.
17. BR 4005, *Manual of Naval Security*, Vol. 1. TNA ADM 234/1134 and Government Security Classifications, April 2014. Cabinet Office.

18 MOD, memorandum, Blelloch to DUS(P) MOD, 6 August 1979, TNA DEFE 24/2122.
19 MOD, memorandum, Pym to Thatcher, MO 13/1/34, 'NATO long range theatre nuclear forces', 5 July 1979. TNA DEFE 25/335.
20 Ultimately, however, the twin track led to the successful Intermediate-Range Nuclear Forces Treaty of 1987; the only arms control treaty to achieve withdrawal of a complete category of nuclear missiles.
21 MOD, loose minute DUS(P) 619/79, 'TNF modernisation – visit by Mr David Aaron', 22 October 1979. TNA PREM 19/14.
22 Cabinet Office. Memorandum, Armstrong (Cabinet Secretary) to Prime Minister, A0940 'Modernisation of NATO's long-range theatre nuclear forces and the replacement of Polaris', 12 December 1979. TNA PREM 19-0159.
23 Cyrus Vance, 'Statement to the press after NATO High Level Group special meeting 12 Dec 79', US State Department (1979). Available: http://usa.usembassy.de/etexts/vance5688.htm [accessed 25 October 2015].
24 Kate Hudson, *CND now more than ever: the story of a peace movement* (Matrix Digital Publishing, 2007), [Kindle edn] location 2192.
25 Missile numbers quoted in: J. Goldhamer, 'The economist's perception of the US–Soviet strategic balance: an update for 1979–1981'. Santa Monica, CA: RAND Corporation,1983, Table 1.
26 Vance, 'Statement to the press'.
27 Cabinet Office. A01003 note of a meeting in the Oval Office, the White House, Washington DC, on Monday 17th December 1979. TNA PREM 19-0159s.
28 In the event the USSR invaded Afghanistan in December 1979 and the US Senate did not ratify SALT II, although both 'sides' did observe its restrictions.
29 Cabinet Office. Memorandum, Wade-Gery (Cabinet Office) to Armstrong (Cabinet Secretary), B05909 'Polaris replacement', 11 February 1980. TNA PREM 19-0159.
30 HC Deb. 24 January 1980. *Hansard*, vol. 977 cols 672–784: 673.
31 *Ibid.*, col. 683.
32 *Ibid.*, col. 686.
33 *Ibid.*, col. 692.
34 *Ibid.*, col. 711.
35 *The Times*, 'More information must be given' (25 January 1980), p. 8.
36 *Daily Express*, 'Maggie's Cold War' (25 January 1980), p. 1.
37 *The Times*, 'Implications of decision to buy Trident' (26 February 1980).

38 N. Cameron, 'Maintaining Britain's nuclear capability', *The Times* (9 May 1980).
39 MOD, memorandum, Pym to Prime Minister, MO 18/1/1 'Polaris successor', 10 June 1980. TNA PREM 19/417.
40 Home Office, unreferenced personal memorandum, Whitelaw (Home Secretary) to Pym (MOD), 'Polaris successor', 11 June 1980. TNA PREM 19/417.
41 FCO, memorandum, Carrington to Prime Minister, PM/80/45 'Polaris successor', 12 June 1980. TNA PREM 19/417.
42 Treasury, unreferenced personal memorandum, 'Polaris successor', 16 June 1980. TNA PREM 19/417.
43 Cabinet Office. Unreferenced memorandum, 'Polaris successor', 17 June 1980. TNA PREM 19/417.
44 MOD, unreferenced memorandum, Whitmore (Cabinet Office) to Norbury (MOD), 'Polaris successor', 7 July 1980. TNA PREM 19/417.
45 FCO, telegram 2517, 14 July 1980. TNA PREM 19/417.
46 John Nott, *Here today gone tomorrow: memoirs of an errant politician* (London: Politico's, 2002), 216.
47 MOD, unreferenced memorandum, Norbury (principal undersecretary at MOD) to Whitmore (Cabinet Office), MO18/1/1 'Polaris successor', 23 June 1980. TNA PREM 19/417.
48 HC Deb. 15 July 1980, statement on the strategic nuclear deterrent. *Hansard*, vol. 988 cols 1235–51: 1236.
49 *Ibid.*, col. 1237.
50 Cabinet Office. OGD 80/23 *The future United Kingdom strategic nuclear deterrent force*. London: HMSO, 1980, para. 33.
51 *Ibid.*, para. 58.
52 *Ibid.*, paras 63–7.
53 Cabinet Office. 'Future of the strategic deterrent', 2 November 1979, para. 6.
54 Defence Committee, *Fourth Report from the Defence Committee Session 1980–81*. London: HMSO, 1981b, para. 1.
55 *Ibid.*, para. 54.
56 Defence Committee, 'Defence Committee alternative draft report dated March 1981'. London: HMSO, 1981a, p. lxiv para. 110(xx).
57 *Ibid.*, p. xl paras 14–16.
58 MOD, personal memorandum, Nott (MOD) to Prime Minister, MO18/1/1 'Trident, public attitudes', 2 February 1981. TNA PREM 19/555.
59 *Ibid.*, para. 4.

60 *Ibid.*, para. 5.
61 *The Times*, 'Mr Nott to fight CND ideas' (20 May 1981).
62 HC Deb. 3 March 1981, Nuclear deterrence. *Hansard*, vol. 1000 cols 137–224: 137.
63 *Ibid.*, col. 144.
64 *Ibid.*, col. 213.
65 Paul Byrne, 'Nuclear weapons and CND', *Parliamentary Affairs*, 51 (1998), 424–34: 426.
66 Although the Party Conference voted in favour of a policy of unilateral nuclear disarmament in 1980 and 1981, the required ⅔ majority was not reached until the 1982 conference; see Hudson, *CND now more than ever* and Byrne, 'Nuclear weapons and CND'.
67 *Daily Express*, 'Foot in "ban the bomb" storm' (27 October 1980), p. 2.
68 *Daily Express*, 'A disarming Michael Foot' (27 October 1980), p. 8.
69 *The Times*, 'CND marches back' (28 October 1980), p. 13.
70 ITV (30 October 1980).
71 *Daily Express*, 'Unlucky break' (30 October 1980), p. 27.
72 *Times*, 'CND marches back', p. 13.
73 Sarah Hipperson, 'Greenham Common women's peace camp' (online: 2000). Available: www.greenhamwpc.org.uk [accessed 17 April 2016].
74 Hudson, *CND now more than ever*, location 2565.
75 Cabinet Office. Memorandum, Cabinet Secretary to Cabinet, OD(80)23 'Defence and Overseas Policy Committee civil home defence', 18 March 1980. TNA CAB 148/190, paras 27 and 28.
76 Cabinet Office. OD(80) 9th meeting of Cabinet Defence Overseas Policy Committee minutes Thursday 20 March 1980. TNA CAB 148/189.
77 Conservative Party, letter Jopling (Chief Whip) to Whitelaw (Home Secretary), unreferenced 9 May 1980. TNA PREM 19/689.
78 Cabinet Office. OD(80) 18th meeting Cabinet Defence and Overseas Policy Committee meeting minutes Tuesday 8 July 1980. TNA CAB 148/189.
79 Home Office, *Protect and survive: civil defence manual of basic training*. London: HMSO, 1950.
80 Home Office, 'Civil Defence handbook no. 10: advising the householder on protection against nuclear attack'. London: HMSO, 1963.
81 Bruce Kent, 'Protest and survive', *History Today*, 49 (1999), 14–16: 14.
82 John Minnion and Philip Bolsover, *The CND story: the first 25 years of CND in the words of the people involved*. London: Allison and Busby, 1983, 89.

83 Hudson, *CND now more than ever*, location 2347.
84 The photographer was Peter Kennard and the Tate reference is T12478.
85 *The Times*, 'CND "puts people in danger"' (18 December 1981).
86 Kent, 'Protect and survive', 15.
87 Peter M. Sandman and JoAnn M. Valenti, 'Scared stiff – or scared into action', *Bulletin of the Atomic Scientists* (1986) 12–16: 12.
88 Deron Overpeck, '"Remember! it's only a movie!" Expectations and receptions of The Day After (1983)', *Historical Journal of Film, Radio and Television*, 32 (2012), 267–92: 282.
89 Louise Lawrence, *Children of the dust*. London: Random House, 1985.
90 Robert Swindells, *Brother in the land*. London: Oxford University Press, 1984.
91 Wheldon to Adam, 31 December 1964.
92 Philip A. G. Sabin, *The Third World War scare in Britain*. Basingstoke: Macmillan Press, 1986, 118.
93 '*Pacem in terris*: Encyclical of Pope John XXIII, 11 April 1963'. Available: www.vatican.va/content/john-xxiii/en/encyclicals/documents/hf_j-xxiii_enc_11041963_pacem.html [accessed 17 October 2019].
94 US National Conference of Catholic Bishops, 'The challenge of peace: God's promise and our response – a pastoral letter on war and peace'. *Origins* (1983): 2. Available: www.usccb.org/issues-and-action/human-life-and-dignity/war-and-peace/nuclear-weapons/upload/statement-the-challenge-of-peace-1983-05-03.pdf [accessed 23 August 2021]. This is a sixty-four-page document of detailed and close argument. Like the International Court of Justice ruling on the legality of the use of nuclear weapons, it is often quoted very selectively to make a given point, to support opposite sides of the same argument.
95 'Nuclear weapons: society, religion and technology project'. Edinburgh: Church and Society Council of the Church of Scotland, 2014. Available: www.churchofscotland.org.uk/data/assets/pdf_file/0003/8895/Nuclear_Weapons_leaflet.pdf [accessed 12 September 2019].
96 Scottish Bishops, Pastoral letter April 1982, repr. Cardinal O'Brien, 'Nuclear weapons – replacing Trident – a Scottish Catholic response', in: Archdiocese of St Andrews and Edinburgh (ed.), 'Justice and peace pamphlet'. Glasgow: Justice and Peace Scotland, 2006.
97 Cabinet Office. Letter, Goodall (Cabinet Office) to FCO, 'Meeting with the Roman Catholic Bishops of England and Wales: nuclear defence issues', 14 January 1983. Margaret Thatcher Foundation Archive: Thatcher MSS (Churchill Archive Centre): THCR 1/4/7 f4.

98 Tanya Ogilvie-White, *On nuclear deterrence: the correspondence of Sir Michael Quinlan*. London: Routledge, 2011, 92. Sir Michael Quinlan is a foundational figure for the study of every aspect of British nuclear policy. Ogilvie-White's curated collection of his correspondence is an invaluable resource for the student of the ethics of nuclear deterrence.
99 MOD, letter, Pym (Private Secretary) to General Synod of the Church of England (General Synod debate on 'The church and the bomb'), 24 February 1983. Margaret Thatcher Foundation Archive: TNA PREM 19/1960.
100 G. Markus, 'Comment'. *New Blackfriars*, 72 (1991), 54–5: 54.
101 'The storm that threatens: statement by the Irish Bishops' Conference on War and Peace in the Nuclear Age', *The Furrow*, 34 (1983), 589–95.
102 HC Deb. 11 March 1982 – Trident missile programme. *Hansard*, vol. 19 cols 975–86: 984–5.
103 Cabinet Office. Memorandum, Armstrong (Cabinet Secretary) to Prime Minister, A09119 'The United Kingdom strategic deterrent; missile processing', MISC 7(82)4, 27 July 1982. TNA PREM 19/0695.
104 Conservative Party, unreferenced memorandum, Ingham (Press Office) to Prime Minister, 'Media relations – stocktaking and looking ahead', 3 August 1982. Margaret Thatcher Foundation Archive.
105 Conservative Party, unreferenced memorandum, Ingham (Press Office) to Butler (Cabinet Office), 'Liaison committee', 9 September 1982. Margaret Thatcher Foundation Archive.
106 Conservative Party, unreferenced 'note of Liaison Committee meeting, Friday 10 September 1982'. Margaret Thatcher Foundation Archive.
107 Conservative Party, unreferenced 'note of Liaison Committee meeting, Wednesday 20 October 1982'. Margaret Thatcher Foundation Archive.
108 HC Deb. 3 March 1981, Nuclear deterrence. *Hansard*, vol. 1000 cols 137–224: 137.
109 Conservative Party, 'Note of Liaison Committee meeting', 20 October 1982, para. 2.
110 HL Deb. 24 March 1983, Defence Secretariat 17 and 19. *Hansard*, vol. 440 cols 1226–30: 1228.
111 *Ibid.*, col. 1229.
112 Cabinet Office. CC(83)14th Conclusions of a meeting of the Cabinet held at 10 Downing Street Thursday 28 April 1983. TNA CAB 128/76/14.
113 Christopher Andrew, *Defence of the realm: the authorised history of MI5*, Kindle edn. London: Penguin, 2009, 676.

114 Joan Ruddock, *Going nowhere: a memoir*. London: Biteback Publishing, 2016, [Kindle edn] location 1020.
115 Stephen Dorril, *The silent conspiracy*. London: Heinemann, 1993, 25.
116 *The Guardian*, 'CND, miners "under MI5 monitoring"' (21 February 1985), p. 1.
117 Gordievsky was a KGB officer in London and acted as a double agent for the British Secret Intelligence Service from 1974 to 1985. See: Andrew, *Defence of the realm*, 674.
118 Ruddock, *Going nowhere*, location 1226.
119 Andrew, *Defence of the realm*, 559.
120 Ruddock, *Going nowhere*, location 1237.
121 *The Guardian*, 'Time for a little self-scrutiny' (10 May 1985).
122 Gerald Kaufmann, 'Thatcher triumphs again' (London: BBC, 1983). Available: http://news.bbc.co.uk/2/hi/uk_news/politics/vote_2005/basics/4393313.stm#issues [accessed 25 October 2015].
123 John Curtice, 'The 1983 election and the nuclear debate'. In: Catherine Marsh and Colin Fraser (eds), *Public opinion and nuclear weapons*. Basingstoke: Macmillan Press, 1989, 51.
124 In 1979, NATO's deployment of Cruise and Pershing II missiles had been agreed as a tactic to force the USSR to the negotiating table, in order to withdraw the entire class of intermediate-range nuclear weapons from Europe.

6

Ethical considerations and wicked issues

In 2016, while making the case for the retention of the UK's independent nuclear deterrent, Defence Secretary Sir Michael Fallon asked 'how would the United States or France respond if we suddenly announced that we were abandoning our nuclear capabilities yet will still expect them to pick up the tab and to put their cities at risk to protect us in a nuclear crisis?'[1] Ethically, and to a great extent strategically, his question poses significant issues.

Chapter three concluded that Churchill's government had gone out of its way to claim the moral high ground during the Second World War and had then masked the true nature of the strategic bombing campaign in order to present this to the public in what seemed a more acceptable light. Chapter two showed that aversion to admitting to a strategy that deliberately involved non-combatant casualties can be seen to have had its roots in the First World War, but it is equally clear that the lexicon to accommodate these concepts was developing in parallel with, or perhaps in reaction to, the realities of 20th-century warfare. Chapters four and five suggested that although subsequent British governments avoided public engagement with the ethical issues associated with nuclear weapons, the individual leaders did feel the weight of the moral burdens they carried.

This chapter will not seek to provide an in-depth analysis of the history of thinking on the morality of nuclear weapons, but will attempt to describe current (2021) thinking in the context of a constantly evolving paradigm. It does, on occasion, lapse into the first person since there is no definitive metric against which to 'measure' ethics. This chapter will conclude with two brief case studies of

The just war tradition – the traditional approach

> There may be dark days ahead, and war can no longer be confined to the battlefield, but we can only do the right as we see the right, and reverently commit our cause to God.[2]

King George VI said this in his radio broadcast to the Empire on 3 September 1939; the day Britain declared war on Nazi Germany. One significant objective of this speech appears to have been to claim the 'moral high ground' and give the British peoples the reassurance that they were doing the 'right' thing. King George addressed the issues in terms of the commonly held tenets of the 'just war' tradition.

He reassured listeners that there was a *just cause* for the war; the whole broadcast was designed to make the British peoples feel that, however much they may have abhorred the need for a further world war, the country had been 'forced into a conflict, for which we are called, with our allies to meet the challenge of a principle which, if it were to prevail, would be fatal to any civilised order in the world' and that 'the freedom of our own country and of the whole British Commonwealth of nations would be in danger'. He emphasised that war was the *last resort*: '[o]ver and over again, we have tried to find a peaceful way out of the differences between ourselves and those who are now our enemies, but it has been in vain'. He was clear that it was being ordered with the *right intent*: '[f]or the sake of all we ourselves hold dear, and of the world order and peace, it is unthinkable that we should refuse to meet the challenge'. War was obviously being declared by a *competent authority*; he was the king – but he couched the whole speech in terms of the collective 'we', implicitly including his democratically elected government in the authority (and also of course the responsibility). In 1939, there was little doubt in government that Britain was ill-prepared for war, but it would hardly have been appropriate for this speech to contain a detailed analysis of the *chance of success*, or the *proportionality* of the ends. His concluding paragraph did give a flavour of the struggle his government was anticipating: '[t]he task will be hard. There may be dark days ahead, and war can no longer be confined to the battlefield.'[3] The whole speech offers an implicit commentary on the priorities of the government at the time: that the Second World War was clearly considered from the Allies' perspective as a just war.

Supreme emergency, dirty hands and nuclear deterrence

As described in chapter three, in 1942–5, the British government did not feel able to describe the mission of RAF Bomber Command accurately to the public, preferring instead to dissemble. Present-day ethicist Alex Bellamy concludes that:

> British military and political leaders did not justify themselves by reference to either the moral tragedy they confronted or the need for special permissions. Significantly, they chose not to do either of these things because they calculated that a significant portion of the British public would oppose the deliberate bombing of German non-combatants and believed that this could undermine domestic support for the war. In turn, this suggests that proponents of the idea of supreme emergency overestimate the extent to which liberal societies are prepared to accept the deliberate killing of non-combatants by their governments, and underestimate the normative force of non-combatant immunity.[4]

Chapters three, four and five suggest that similar reticence has been prevalent throughout British consideration of acts of war that may cause significant non-combatant casualties, and in particular the public presentation of nuclear deterrence policy. At the start of the Second World War, King George VI set a high ethical standard for the British entry into, and conduct of, the war; standards that Roosevelt's 1939 speech suggests he clearly hoped combatants would sustain.[5] A democratic state requires the support of the population in order to commit to something like a total war – and King George's address made clear that he was under no illusions about how 'total' the forthcoming conflict was going to be. Faced with a nuclear war, a similar – if not greater – level of commitment would be necessary from the population, and the government of the day would be faced with the task of preparing the public, materially and conceptually.

The public of 1939 understood what war meant – many could remember the First World War and most families could number family members who had been lost in that conflict. The modern public has almost no experience or understanding of war, other than as something for which professional armed forces deploy elsewhere. Experience of the management of the coronavirus pandemic in 2020/21 suggests that preparing a British 21st-century public for a war with significant domestic ramifications would be extremely demanding, not least in the justification of the measures being taken, the use and publication of expert advice and opinion, and government engagement with dissenting views.

This chapter aims to investigate what moral factors statesmen and those involved in the nuclear deterrence mission consider. In particular, recourse to the 'supreme emergency argument' will be tested in the context of contemporary ethical arguments although, as demonstrated in the historical analyses in chapters two to five, there is little historical evidence of a direct link between contemporary ethical thought and decisions on UK nuclear deterrence policy; which is why, as a rare public indication of official understanding of how nuclear deterrence 'works', Michael Fallon's statement about holding cities at risk might be seen as ethically problematic. I do tend to infer regular links between the ethics and the public presentation of that policy, though.

Almost without exception, contemporary Western commentaries on the ethical aspects of warfare allude to the just war tradition. The use of the term 'tradition' is a careful one; just war thinking is too loose a concept to merit description as either a school of thought, a theory or a paradigm. But there is clearly a thread of ethical thinking related to the conduct of violent relations between political entities that seems to merit some form of definition. Bellamy describes it as 'a protracted normative conversation about war that has crystallised around a number of principles'.[6] Within this tradition or conversation, many writers on the ethics of war incorporate references to historical figures (reaching back as far as

St Augustine) who have contributed to the evolution of the tradition, perhaps to demonstrate a degree of philosophical antecedence or authority.

Modern moral philosophy is a factor for contemporary statesmen, officials and military personnel, but it seems to be only a factor, and those more intimately involved in ethics naturally feel that they would prefer to see it take a more central, if not pre-eminent role. There appears to be a difference in common morality that is insufficiently represented in the revisionist canon of ethics but which, however, is present in the thinking of those with responsibility for waging war. Accepting that those responsible for Britain's armed forces and nuclear deterrent are neither immoral, amoral nor simply hypocrites, logically when formulating a moral framework they must work to an ethical approach that considers factors different from those in common currency among philosophers. Such a framework would seem to need to be consequentialist in its derivation and might appear ruthless and therefore very easy to pillory in terms of theories of rights-based morals and revisionist critiques of the just war tradition.

I do not intend to provide a history or analysis of the derivation of the tradition. However, the relative importance of each of the parameters commonly considered to inform it has varied in contextual relevance, and this transient salience is an important factor when considering the contemporary debates within the analysis of the ethics of the use of force, and will therefore be considered when relevant. Johnson specifically analyses, and contributes to, the evolution of thinking on the ethics of war:

> Both just war and pacifist thought in Western culture have developed as historical traditions shaped by diverse influences ... Exactly how these traditions have developed historically is fundamentally important for understanding them and drawing out their meaning in the contemporary context.[7]

Bellamy suggests that:

> The just war tradition fulfils two roles. It provides a common language that actors can use to legitimise recourse to force and the conduct of war and that others can use to evaluate those claims. It can also inhibit actions that cannot be legitimated.[8]

The suggestion of a 'common language' will be further examined in this chapter. Fifty years ago, Quade argued that 'the principles of just war become operative only after the classic political question is answered: who should do the judging?'[9] Clearly, for Augustine writing in the 5th century CE, this was simple; the actor was the judge and should act morally, and would face a divine accounting eventually. This fundamental question has yet to be answered definitively for the secular actor, and it remains just as important. The nature of person judging the morality of an action changes the nature and purpose of the codification of the moral thinking; is 'just war' thinking simply a tool for policy-makers, or an analytical framework for subsequent philosophical debate, or should it be both?

This is not an esoteric distinction; it feeds a number of subordinate, but significant, ethical issues. If the person who should judge the morality of an action is the actor themself, how do they ensure they have sufficient information about the ramifications of the action in order to make that moral judgement? How can this be credible when it comes to using nuclear weapons? Can a subordinate assume that their cause is just because their superiors have assured them that it is so? Finally, if the appropriate judge is someone other than the moral actor, how can they apply an authoritative set of ethical criteria?

Writing in 1978, while the Cold War was at its chilliest, ethicist James Childress wondered:

> What degree of certitude should policy makers and citizens have about the justice/justification of a particular war? Should they be convinced that the preponderance of the evidence indicates that the war is just/justified according to the above criteria? Or should they be convinced beyond a reasonable doubt?[10]

Childress does not actually answer this question. King George VI was clear that he believed that 'we can only do the right as we see the right' and Browne, writing in 1945, seemed to agree: '[a]n action is moral, we shall say, when the agent judges that the act is right and the action would not have occurred if the agent had not so judged. Thus there must be a desire or intention to do what is right as such.'[11]

Both Childress and Browne seem to imply that the agent of the moral action is the appropriate authority to determine whether it is just. Clearly, this was the view held by King George VI. Whilst common, this is not a universal view; ethicist Michael Walzer argued that 'the moral reality of war is not fixed by the actual activities of soldiers but by the opinions of mankind. That means, in part, that it is fixed by the activity of philosophers, lawyers, publicists of all sorts.'[12] This suggests that he considered that the just war tradition could be an appropriate guide for the actor, but instead of an Augustinian divine accounting, Walzer's actor would be appropriately held to account by philosophers, lawyers and publicists. He continued: '[i]n moral life, ignorance isn't all that common; dishonesty is far more so. Even those soldiers and statesmen who don't feel the agony of a problematic decision generally know that they should feel it ... If they don't, they lie about it.'[13]

Michael Quinlan and Charles Guthrie, experienced and senior figures in Britain's defence establishment during the end of the Cold War, have subsequently argued '[decisions to go to war] entail taking very serious decisions on the basis of estimates of complex futures with wide margins of uncertainty and as a result much scope – often on both sides of a conflict – for different perceptions and judgements about where justice and prudence point'.[14] In a dwindling minority among theorists, James Turner Johnson argues vigorously that 'the decision whether to resort to use of armed force is properly the responsibility and right of those in positions of supreme political authority in a society, not that of moralists'.[15] Neither are moralists, or publicists, omniscient:

> The difficulty of knowing what is right has been made more apparent by the recent shift away from absolutist types of ethical theory toward teleological, or consequential, types. For if the rightness of an act depends on its leading to consequences of a certain sort, knowledge of its rightness requires knowledge of its consequences, and this requires experience. The relevant consequences may in fact be so numerous and far-flung that anticipation of more than a small fraction of them is beyond the power of the wisest man.[16]

Ethical considerations and wicked issues   151

This difficulty in predicting consequences is echoed by Fisher:

> A more serious criticism is that the kind of god-like calculation required to draw up such a balance sheet before the onset of war is beyond the wit of man, given the uncertainty and unpredictability of war and the incommensurability of the values to be thus balanced. What probabilities are to be assigned to the possible outcomes?[17]

Obviously, these criticisms are valid only from a consequentialist perspective. This butts up against contemporary ethicists such as David Rodin and Jeff McMahan who regard the rights of individual human beings as entailing *prima facie* duties and obligations, and who do not attribute any inherent moral value to political communities, and states in particular.[18] Thus the rights to life of individuals cannot be abrogated under any circumstances, including defence of a community. These writers' efforts to identify a catechism to encapsulate this, however, have so far failed to convince the mainstream of political thought and as a result their serious contributions to the understanding of the factors that statesmen should be taking into account are condemned to the sidelines; functional pacifism, as Johnson put it.[19]

This does, however, have a pernicious side-effect; it threatens to hijack the lexicon of the just war tradition. If the right of national self-defence is rendered meaningless in the moral vocabulary offered by modern ethics, then the language being used inhibits the thought processes available to those involved. The reduction of the richly complex vocabulary of the ethics of international relations to that of pacifism introduces intolerable inhibitions on the moral use of force. Legitimate regimes would struggle to justify the use of force in almost every circumstance and, ironically, the way would then be easier for immoral regimes to use force without opposition. It appears that the most important thing about a public moral discourse is not that it has a single normative solution, but that the discourse exists. Statesmen must feel obliged to justify their actions, and ethicists must be able to challenge them; and the lexicon of the just war tradition provides the language and paradigm for that exchange; it is diminished at our peril.[20] The just war tradition does not constitute a normative basis on which international relations are conducted, but it does provide the words we use to talk about them. It is, perhaps, in interpretation of some of those terms that the ethical paradigm within which I (and, I suspect, my contemporaries, senior officers, officials and politicians) justify nuclear deterrence to ourselves differs from Rodin and McMahan; although we all use the same words, the words' connotations differ. This therefore makes it more difficult to engage in public on issues like strategic bombing and nuclear deterrence.

During the Cold War the deterrence strategies of both the USSR and NATO were based on the premise that they could fight, and win, a nuclear war. Each side had tens of thousands of nuclear weapons which would be used for their military effect; they would be used as explosives to destroy military targets and to achieve other 'military effects'. The nuclear element of the NATO plan was known as the Single Integrated Operating Plan; it identified more than fifty military targets in Moscow alone and authority was delegated to battlefield commanders for the release of nuclear weapons for their tactical (military) effect. Ethically, this differed little from the strategy of the Second World War and previous conflicts; destruction of military targets would cause an accumulation of military effects until the defeat or capitulation of the other side. In itself, this capability to fight and win a nuclear war was perceived as an effective deterrent. The release of tactical nuclear weapons was delegated through the NATO command chain to battlefield commanders.

As discussed in chapter four, the ethical justification for this strategy was derived from the same doctrine of double effect as the (still hotly contested) strategic bombing campaign of the Second World War.[21] In the 1980s, this entirely predictable 'collateral damage' ran to hundreds of millions of casualties, and the prevailing public discourse was dominated by fear (the emotive essence of both the official Protect and survive and the CND Protest and survive campaigns) and also ethics. The 1983 US National Conference of Catholic Bishops' pastoral letter 'The challenge of peace' was written after very extensive consultation between the conference and the US nuclear deterrence community. The bishops drew the highly nuanced conclusion '[i]n current conditions "deterrence" based on balance, certainly not as an end in itself but as a step on the way toward a progressive disarmament, may still be judged morally acceptable'.[22]

I struggle to understand how Cold War deterrence strategies based on nuclear warfighting could be deemed morally acceptable since the proportional costs would be 'colossal', as Walzer put it in 1977.[23] Deterrence strategies have changed in the last forty years, though the ethical discourse does not appear to have evolved to keep pace. Precisely because of the limitations of proportionality, no political objective can justify nuclear aggression in order to achieve it; deterrence of war, however, is a morally legitimate objective, and deterrence is what nuclear weapons are 'for'. No purely military effect is ethically proportionate to the level of non-combatant casualties that the use of nuclear weapons would entail. In NATO nuclear deterrence exercises, a frequent debate is often 'could this target not be struck with conventional weapons instead?' The answer is usually 'Yes, but the target doesn't need the nuclear weapon; the crisis does'. Modern nuclear deterrence strategies are not predicated on using nuclear weapons for military effect but for the political effect which is the imposition, or reimposition, of deterrence; not the destruction of a given military target. Where national survival or critical national interests are threatened – a 'supreme emergency' – the just war tradition's recalculation of the proportionality criterion might render the use of nuclear weapons as tools of deterrence if not legitimate, at least excusable.

The premise of British nuclear deterrence in the 21st century is that nuclear force would only be used in the context of what Churchill called the supreme emergency:

> Our defeat would mean an age of barbaric violence ... we have a right, indeed are bound in duty, to abrogate for a space some of the conventions of the very laws we seek to consolidate and reaffirm ... The letter of the law must not in supreme emergency obstruct those who are charged with its protection and enforcement ... Humanity, rather than legality, must be our guide.[24]

Walzer put this into the ethical context:

> In an emergency, neutral rights can be overridden, and when we override them we make no claim that they have been diminished, weakened, or lost. They have to be overridden, as I have already said, precisely because they are still there, in full force, obstacles to some great (necessary) triumph for mankind.[25]

The defence of liberal democracy in western Europe against Nazism was a political effect justifying abrogation of neutral states' rights under the 'supreme emergency' condition, which the simple military effect of preventing Wehrmacht forces landing in Norway might not. In the context of the 'supreme emergency', the constraints on actions still exist, but they are temporarily rendered proportionate directly to the political effect they are intended to achieve, not their immediate military effect.

In contemplating the use of nuclear weapons for deterrence, I find myself aligned with this 'dirty hands' argument, the essence of which is that a person can be faced with a situation where they have two options, both of which appear morally reprehensible, but one of which must be carried out. The military euphemism for this is 'we have two options; bad or worse'. This argument pivots on consequentialist morals. Can the ends sometimes excuse the means? The moral actor must appreciate that while the ends might excuse their actions, they do not justify them: their actions mean that they are not innocent. However, that is not to say they are guilty:

> Why shouldn't he have feelings like those of St Augustine's melancholy soldier, who understood both that his war was just and that killing, even in a just war, is a terrible thing to do? … Here dirty hands are a kind of impurity or unworthiness, which is not the same as guilt, though it is closely related to it.[26]

As considered above, this is unlikely to find approval amongst a jury of ethicists. Walzer conceded that even *in extremis*:

> [the rights possessed by individual human beings regardless of circumstance] cannot be eroded or undercut; nothing diminishes them; they are still standing at the very moment they are overridden: that is why they have to be overridden. Hence breaking the rules is always a hard matter, and the soldier or statesman who does so must be prepared to accept the moral consequences and the burden of guilt that his action entails.[27]

In the dirty hands context, Walzer wrote:

> When rules are overridden, we do not talk or act as if they had been set aside, canceled, or annulled. They still stand and have this much

effect at least: that we know we have done something wrong even if what we have done was also the best thing to do on the whole in the circumstances.[28]

Rodin does not accept either the supreme emergency or the dirty hands argument: '[a]ccording to Walzer ... a community is permitted to violate the most basic *in bello* norms if doing so will enable it to avoid destruction at the hands of a military aggressor'.[29] Bellamy describes it as 'when a state confronts an opponent who threatens annihilation, it can be morally legitimate to violate one of the cardinal rules of the war convention – the principle of non-combatant immunity'.[30] Just like the dirty hands argument, the concept of supreme emergency recognises that it does not justify, merely excuses: '[i]t is the acknowledgment of rights that ... forces us to realize that the destruction of the innocent, whatever its purposes, is a kind of blasphemy against our deepest moral commitments. (This is true even in a supreme emergency, when we cannot do anything else.)'[31]

There is a further complication; deterrence is the threat of the use of force in order to prevent an adversary taking a course of action which would require the use of force to counter. Nuclear deterrence is not about blowing up cities, it is about interfering with the decision-making process of an adversary who is concerned that you might blow their cities up if their actions forced you to. Cold War philosopher Anthony Kenny argued: '[i]f someone involved says he would pull the trigger if deterrence fails, then I can only tell him, quite soberly, that he is a man with murder in his heart'.[32] Having made that decision myself, I do not agree; on one level, this is a person prepared to kill in war, but actually this is a person who, by being prepared to do so, helps to prevent war; and the metaphysical action must be weighed against its tangible outcomes.

French strategic commentator Bruno Tertrais suggests that no major power conflict has taken place in nearly seventy years; there has been no direct military conflict between nuclear powers; no nuclear-armed country has been invaded; and no country covered by a nuclear guarantee has been attacked, conventionally or otherwise.[33] He makes in detail a compelling case that nuclear weapons have suppressed major conflict between the great powers of the world, and have not simply deterred nuclear conflict. Of course, myriad other commentators disagree, but it is difficult to avoid the conclusion that nuclear deterrence has at least contributed to the prevention of war between the major powers for seventy-five years: the metaphysical evil of threatening violence is far outweighed by the good of preserving the (albeit imperfect) peace: in just war terms; the risk of non-combatant casualties is proportionate to the deterrence of war.

Prior to assuming command of HMS *Vengeance* in 2003, I considered the implications of the role as fully as I was able. As I suggested in the introduction, I sought to understand the absence of an officially endorsed position, and I have continued to develop my own view of the ethics and the strategy of 21st-century nuclear deterrence ever since. The dirty hands and supreme emergency arguments described above best fit my experience, and inform my commitment to a strategy of deterrence while recognising the repugnance inherent within it. This is a thought process every commanding officer (CO) must go through in their preparations for command and it boils down to one issue: the level of trust (effectively faith) that the CO must place in the national firing chain; the mechanism by which the order to launch is formulated and transmitted must be absolute.

Only the Prime Minister, or their nominated nuclear deputy in the event of their incapacity, can authorise launch of British nuclear weapons. Once that decision is made, the Prime Ministerial Directive – the authorisation to launch – also requires the Chief of Defence Staff (CDS) to concur and to add their coded contribution, without which the directive would be meaningless. (Peter Hennessy speculates whether the CDS or senior civil servants would be able to delay or prevent a launch if they believed that the PM had gone 'bananas … at a period of high international tension'. His research with Lord Guthrie (CDS 1997–2001) and Sir Frank Cooper (permanent under-secretary to the MOD 1976–82) suggested that they would.)[34] The Prime Ministerial Directive is transmitted to Commander Task Force 345 in an underground bunker at Northwood on the north-west of London where it is re-encoded into the national fire control message, which is transmitted directly to the submarine; although it is transmitted on military networks, there is no intervening military level of command. Uniquely in the armed forces, the CO of an SSBN must simply obey an order to launch and they have to make that decision prior to taking command; based on their faith in the rigour of this command chain. This is not 'following orders' regardless of their legitimacy in the sense of blind obedience; in essence, the CO of the SSBN has already carefully thought through this issue and decided to follow the order to launch before assuming command.

My predecessors, successors and I, commanding the continuous deterrence patrols which have provided the UK with national strategic nuclear deterrence since 1969 have always done so regardless of which party is in government. We (and I use the pronoun carefully) always had reason to be confident that the order to launch is not capricious nor subject to any interests other than extreme national interest, and that the appropriate expert scrutiny – political, legal and ethical – had been applied to any decision to launch before the order is sent. Our only recourse for that assurance was a fundamental level of trust in the individual originating that order; and that senior military and civil service figures would have been intimately involved in deterrence decision-making processes. Every SSBN CO must have absolute faith in the integrity of the individual giving the order to launch.

As considered briefly in this section, one limitation of moral discourse is that it inhibits its own parameters by its lexicon. Modern Western philosophy is almost exclusively derived through the lens of the Enlightenment and the focus on the rights of the individual; Rodin and McMahan regard the world as a collection of autonomous individuals, each with absolute rights, and they formulate philosophical arguments accordingly. Psychologist Jonathan Haidt suggests that 'philosophers since Kant and Mill have mostly generated moral systems that are individualistic, rule-based, and universalist. That's the morality you need to govern a society of autonomous individuals.'[35] He goes on to suggest that there exist factors that legitimately inform moral decisions other than those which informed Kant and Mill and subsequently much post-Enlightenment, rights-based philosophy.

'Who judges' therefore is important, because it changes the nature of the thinking behind just war ethics. Policy must be made in real time, with limited information and some assumptions; does the statesperson act in accordance with their own conscience, or in accordance with what they believe will meet the approval of subsequent analysts? And does this then have an effect on their actions? Should it? This tension pervades every strategic decision. The 'jury' is also important in this context because if the jury has an agenda – and the just war tradition has always been susceptible to selective use – it can be used to devastating effect to make decisions (which might have been reasonable and well-considered in the circumstances) appear utterly immoral in hindsight where the 'jury' and the statesperson consider the factors differently. The British public could be seen as one contemporary, real-time 'jury' for the British government.

This chapter started with a review of King George VI's speech of September 1939, considering the ethical tenets it invoked to persuade and encourage the British public as they faced the second major war in their lifetimes. Despite the moralising, rhetorical flourishes and public pronouncements, the British strategy for area bombing was substantially that which the government denied; terror bombing. There are ethical arguments that can be made to excuse this, but there was clearly unease within the British government about how those arguments would stand up in public, even during wartime. In the event, the public acquiesced in a convenient fiction; that non-combatant casualties were the unintended consequences of attacks on military targets.

Chapters four and five suggest that successive governments have simply avoided public discussion of this nature on nuclear deterrence policy, not so much because the policy itself is morally reprehensible, but because it is difficult to present in terms which unambiguously appear 'good' to the public. Plato's ring of Gyges parable still applies;[36] experience of the coronavirus pandemic does not suggest that the public is prepared for nuanced engagement on nuclear deterrence, and certainly does not support the view that the government could make the argument with any sophistication. The next section will consider two alternative models for government engagement. Neither is nuclear, but both issues present alternative models for engagement in connection with demanding ethical matters which elicited considerable public unease.

Wicked issues

It is the lot of government to face 'wicked' issues. Originally a term coined in the social policy environment in the USA, a wicked issue refers 'to that class of social system problems which are ill-formulated, where the information is confusing, where there are many clients and decision makers with conflicting values, and where the ramifications in the whole system are thoroughly confusing ... where proposed "solutions" often turn out to be worse than the symptoms'.[37] Use of this term has expanded beyond the social policy environment and I suggest that the issue of nuclear deterrence policy is just such a wicked issue.

The increasing power of science to affect the world in which we live poses distinct ethical questions which often fit precisely into this definition of wicked issues. The balance between the threats and benefits of scientific innovation in issues as widely spread as artificial intelligence, energy production and climate change, use of animal subjects in research on human diseases, the teaching of evolution in religious schools, nuclear deterrence, and genetic analysis or modification pose some of the key ethical issues facing modern society. One commentator suggests:

> What has happened is that the speed of technological change has outstripped our ability to adjust our social life and ideas. It is clear that technology is forcing us to confront some of our most dearly treasured ideas because it has devised means of doing things that we would never countenance if it were proposed that they were undertaken without the benefit of technological assistance, as most them can be.[38]

Or, more bluntly: '[w]e have created a Star Wars civilization, with Stone Age emotions, medieval institutions, and godlike technology'.[39] The core of these issues tends to be at the heart of modern ethical debate; are there perpetual, absolute moral values which should preclude certain activities, even if those activities promise huge consequential benefits? 'Scientific advances often create "policy vacuums" or situations that demand choices. But the right path is seldom clear.'[40] It falls to government to make those choices, and they are potentially the most 'wicked' choices of all.

In making difficult policy choices, government always has different options. The default tends to be to avoid the issue altogether until the problem is so pressing that something must be done. This salience is often dictated by public engagement and media coverage or, occasionally, particularly effective lobbying in Parliament. Once a wicked ethical issue is in the public domain, there is no shortage of 'experts' prepared to argue for one solution or another, and the most vociferous of these tend to be the more dogmatic. Thus, by the time government becomes engaged publicly in resolution of a wicked issue, there are often already well-established and entrenched positions, all of which will have an element of credibility sustained by selective interpretation and presentation of evidence, and a train of highly opinionated supporters. It is the government task to set bounds to the issue, identify and consider the advice of genuine experts and make decisions on behalf of the society it serves.

This study does not seek to investigate ethical, and in particular moral, positions *per se*, nor to pronounce judgement on government decisions, but simply to consider how those decisions have been made in one of the longest-standing and most wicked ongoing issues; nuclear deterrence policy. In doing so, it has considered British government engagement with the public at key points in the development of British nuclear deterrence policy but, in order to provide a contextual framework for this analysis, this chapter will consider the treatment of two other wicked issues created by scientific advances in highly emotive areas: human embryology and fertilisation; and genetically modified crops. The treatment of each of these issues has been very different, but the scope and symptoms of the 'wickedness' of the issues were similar.

Human embryology and fertilisation

On 7 December 1978, a top secret report titled 'Factors relating to the further consideration of the future of the United Kingdom deterrent' was submitted by the Cabinet Office to the Chancellor, Foreign Secretary and Defence Secretary.[41] The remainder of the Cabinet did not know of its existence. The Duff-Mason Report, as

Ethical considerations and wicked issues 161

it became known, contributed substantially to the decision-making process that culminated in 1982 with the Thatcher government decision to procure Trident from the USA. The process that led to that decision was considered in detail in chapter five. Another 'wicked' decision was playing out during exactly the same period.

On 25 July 1978, baby Louise was born to parents Lesley and John Brown in Oldham Hospital; the first human being to be conceived outside the womb using an experimental technique to overcome infertility, *in vitro* fertilisation (IVF). The event caused an immediate media sensation, and a flurry of strongly opinionated moralistic posturing. On the same day, the government responded to a Parliamentary question asking 'what control is required over research involving human fertilisation *in vitro* and embryo transplants?'[42] and whether the ethical and social aspects should be reviewed, by stating that '[t]he techniques of human *in vitro* fertilisation and embryo transfer do not involve genetic manipulation in any way ... the Medical Research Council ... would not support research in these fields until there was satisfactory evidence from work with animals of the safety of the techniques'.[43] Of course, this did not address the question asked.

The Medical Research Council (MRC) had been aware of the development of IVF since 1971, but had not become involved either in supporting this life-changing medical procedure, or in inhibiting or supervising the development of the morally challenging genetic process it involved. After Louise Brown's birth, the MRC established an 'Advisory Group to Review Policy on Research on *In-Vitro* Fertilisation and Embryo Transfer in Humans'. Its first meeting, in 1979, recorded '[i]n 1971, an application from Mr P. G. Steptoe and Dr R. G. Edwards for long-term support of a programme of studies on human reproduction had been declined because of serious doubts about the ethical aspects of some of the proposed investigations'.[44] Having identified and expressed to itself serious ethical doubts, the MRC seems to have failed to identify that the issue at hand was not whether the research should be funded, but whether it should be authorised.

By 1982, the British Medical Association, the Royal College of Obstetricians and Gynaecologists, the Council of Churches and the Council for Science and Society had all established working parties in this field. Each of these bodies may have asserted authority within its own remit, but this 'wicked issue' overspilled their areas of expertise and prerogative and none was sufficiently broadly based or sufficiently representative to be regarded as a source of authoritative advice to government. The Department of Education and Science advised the MRC:

> These further developments raise difficult moral and legal questions – the rights of a child *vis-à-vis* its genetic and its biological parent; the responsibilities and liabilities of those handling human embryos; the circumstances in which an embryo developed *in vitro* might be kept or destroyed; and so on. Public concern about these issues has been focussed by recent reports in the press and on television, and there have been repeated calls both inside and outside Parliament, for a Government enquiry.[45]

In July 1982, the Department of Health and Social Security (DHSS) announced a government inquiry into human fertilisation

> to consider recent and potential developments in medicine and science related to human fertilisation and embryology; to consider what policies and safeguards should be applied, including consideration of the social, ethical and legal implications of these developments, and to make recommendations.[46]

Announcing the inquiry, the Secretary of State for Social Services continued: '[m]embership will be broad based and includes, as well as doctors and lawyers, other relevant professionals and those with experience in family policy and the child care fields; other "lay" and religious viewpoints will also be represented …'[47] There had been no discussion of the impending enquiry at the MRC Advisory Group meeting on 13 May 1982, suggesting that this was not necessarily a government-wide initiative and there is no mention of the inquiry at all in Cabinet minutes prior to its establishment although, to be fair, the Cabinet was distracted by the Falklands War at the time.

The surviving files do not give a good indication of the government rationale for establishing the inquiry, although there is enough to determine that it was not at all universally popular in Whitehall. The best source of information on the establishment of the inquiry is actually the report of (what became) the Warnock Inquiry itself:

it was our task to attempt to discover the public good, in the widest sense, and to make recommendations in the light of that ... we had, in the words of one philosopher, to adopt 'a steady and general point of view'. So, to this end, we have attempted in what follows to argue in favour of those positions which we have adopted, and to give due weight to the counter-arguments, where they exist.[48]

Warnock and her committee identified that one of the major intangible imperatives for the enquiry was the anxiety generated in the public mind by the developments in scientific and medical techniques. The committee recognised from the outset that, in a pluralistic society, one particular set of principles could never be completely accepted by everyone:

> We recognised that within society there is a multiplicity of views on the issues before the Inquiry. We therefore decided to seek evidence from as many organisations, reflecting as many different perspectives, as possible ... But even with submissions from so many organisations we have to record with regret that we did not receive evidence from as wide a range of minority and special interest groups as we would have liked, despite our best endeavours.[49]

Analysis of the specific issues under consideration is not pertinent here, but a brief consideration of the bodies submitting evidence is instructive.

The list of bodies and individuals providing evidence is considerable and includes twenty-three universities, twelve Royal Colleges, the High Court (Family Division), bodies representing all of the established religions in the UK, more than thirty local authorities, local and national groups with a particular interest in the related issues, charities, individuals and the Trades Union Congress.[50] The MRC, already castigated in Parliament for its failure to identify the larger issues associated with human embryology, was one of the first bodies invited to present expert evidence to the Warnock inquiry.[51] Despite having been silent on the issue previously, the MRC promptly published 'a statement on research relevant to human fertilisation and embryology in the *British Medical Journal* of 20th November'.[52] Subsequent internal correspondence between the secretary of the Warnock inquiry and the chair of the MRC suggests that the relationship between the MRC and the Warnock inquiry was not entirely harmonious; in a draft of a 'stock' DHSS response to letters from the public on the issue, Dr Rashbass (chair of the MRC) struck out by hand two clauses which had read '[i]t would therefore be inappropriate for me to say anything at this point in time which might be seen as prejudicing the outcome of the inquiry under Mrs Warnock's chairmanship' and 'while I in no way wish to prejudice the inquiry's work'.[53] Each of these stock phrases is used routinely within Whitehall to stress interdepartmental coherence, especially where there is none. That this section was regarded as necessary by the Warnock inquiry is indicative of interdepartmental tensions, but that it was struck out by the MRC is, perhaps, more so.

The MRC publication of its evidence prior to presenting this to the Warnock inquiry set a precedent which was followed by a number of other bodies which had hitherto been silent on the issue: the Royal College of Obstetricians and Gynaecologists and the Royal Society each published pamphlets outlining their position, prior to presenting to the inquiry. Each of these publications was reported in the press at the time, pre-empting the Warnock inquiry report. If the MRC and others attempted to influence the public debate prior to the publication of the Warnock Report, they probably succeeded, but the report had the last word. By addressing the issues that the evidence had raised in a measured and holistic fashion, the Warnock inquiry took the rhetorical 'heat' out of the issue and enabled a genuinely informed and rational discussion in the media. Although the public was not spared headlines such as 'could my husband be my brother?'[54] it was also treated to better considered, more balanced discussions such as one on Radio Four's flagship current affairs programme *Analysis*.

In her subsequent analysis of the role of the inquiry, Baroness Warnock recognised the constraints and, in articulating them, defined the crux of the matter under consideration here. 'One of the obvious difficulties is to establish what the moral sentiments of the public actually are. In a pluralistic society, such as ours, there is certain to be a wide diversity of views.'[55] This is a generic observation applicable to any ethical question, and when significant moral factors are involved, the difficulties are exacerbated:

The more certain people are of the correctness of their views, as a rule, the more vocal they are. It tends to be the hard-liners, in whichever direction, who tell their views abroad. And so there is a danger that 'public opinion' may come to be identified not with the views of the relatively confused, relatively open-minded majority, but with the views of the committed and the fanatical.[56]

However, the reality has never been so pure, and public pressure, of the relatively confused or the fanatical, has long had an influence on government policy on ethically challenging issues.

In the early 19th century, an increasing demand for human corpses for anatomical study and surgical training created an underground support industry of grave-robbers: '[w]ithout the bodies, which were illegally obtained through grave robbers, the work of many anatomists would not have been possible'.[57] In 1828, William Burke and William Hare bypassed the necessity of grave-robbery by simply murdering ten people and delivering their bodies to the famous anatomist Doctor Robert Knox. Burke was hanged. With deliberate bureaucratic irony, his body was dissected and his skeleton remains on display in the Medical School at Edinburgh University. *The Lancet* reported at the time, 'Government is already in a great degree, responsible for the crime which it has fostered by its negligence, and even encouraged by a system of forbearance'.[58] Only once the situation was publicised did government act. 'The Anatomy Act of 1832 provided an adequate supply of bodies for the teaching of anatomy, gradually putting out of business resurrectionists: the end stooped to justify the means.'[59]

Warnock further elaborated the constraints on her committee:

The role of an Inquiry such as ours can only be to try to get it right, and above all to consider the moral arguments on each side, such as they are, and to set them out with clarity. This will help Ministers to make whatever case they decide to make, with a view to persuading Parliament.[60] [She rather wryly observed that] it is probable that Ministers hoped for more from us than they got. They may have hoped for a solution to the problems, and a clear unanimous voice explaining what was right and what was wrong. In this we failed, and rightly so.[61]

Warnock's point here is significant – she regarded the function of the inquiry not to determine the 'right' answer, but to consider and present all aspects of the wicked issues. She further described the need to '... sift through and sort out the facts that are relevant to decision making. But it has to be recognised that there is no such thing as a moral expert',[62] and even if there were, the balance of deontological factors in each individual is unique, so no two people ever see eye to eye on every issue. Lippmann argued nearly 100 years ago that the legitimacy of an expert 'depends upon separating himself from those who make the decisions, upon not caring, in his expert self, what decision is made',[63] and Pielke echoes this in a more modern context: '[b]ecause scientific results always have some degree of uncertainty and a range of means is typically available to achieve particular objectives, the task of political advocacy necessarily involves considerations that go well beyond science'.[64]

But, as Warnock argues, 'the only reason to call such people "experts" is so that their conclusions may be accepted as authoritative without question. Other people, both Ministers who have sought their advice and society at large, must be prepared to say "the experts have decided that this and that are right; and we must go along with it"'.[65] This is particularly pertinent for wicked issues, which often have powerful deontological considerations: '[n]ow in matters of life and death, matters of birth and of the family with which we were concerned in the Report, no one is going to give up his beliefs without a struggle. No one is going to accept what someone else thinks right just because he is told he should.'[66] Warnock touches on two issues common to government handling of complex ethical issues; a tendency to defer to experts (so long as the expert opinion matches the government inclination) and the position and definition of experts. This is true of the Warnock inquiry, and many other present-day wicked issues where a government inquiry is established in order to provide 'expert' advice to ministers.

After the report of the inquiry in July 1984, the government established an interim licensing authority to regulate work on *in vitro* fertilisation until legislation could be introduced based on the inquiry's report. There were two unsuccessful challenges to the inquiry's recommendations in Parliament: in 1985 the 'Unborn Children Protection Bill' sought to prohibit embryo research and limit artificial insemination; and a similar bill was presented in 1986. The Human Fertilisation and Embryology Act was passed in 1990 and came into force, establishing the Human Fertilisation and Embryology Authority, in 1991. Since then, a number of statutes of limited scope have been introduced to manage embryology, but none have elicited the public interest that the original inquiry generated.

In 1978, any government action, including inaction, would have been vilified by one or other lobby within this particular wicked issue. Even after the establishment of the Warnock Committee various lobbies, including government bodies, attempted to influence the public debate prior to the publication of the inquiry's report by publishing their evidence to the inquiry, effectively seeking to influence the 'narrative' in the public domain for a while. This clearly resulted in short periods of criticism for the government. But in the longer run the inquiry balanced the opposing views of the various lobbies, and delivered an objective assessment of the risks associated with all of the issues. This allowed the government to remove the highly emotive rhetoric characteristic of many of these lobbies, and to legislate in an environment that was much more dispassionate and representative of a genuinely well-informed public debate; a circumstance sadly lacking on occasion elsewhere.

Genetically modified (GM) crops

The advent of genetic engineering has allowed the modification of crops to provide greater yields or resistance to pests thus enhancing productivity. However, during the late 1990s food scares and lurid media headlines about 'Frankenstein foods'[67] or 'mutant porkies'[68] fuelled British public concern about the safety of these techniques. Government figures indicated that in 2002, '68% of UK adults claimed to be very or fairly concerned about food safety issues'.[69] As with *in vitro* fertilisation, it was the public perception of this wicked issue that drove government action. The Agriculture and Environment Biotechnology Commission (AEBC) was established in 2000 to 'look at current and future developments in biotechnology which have implications for agriculture and the environment, and to advise the Government on their ethical and social implications and their public acceptability'.[70]

It is illuminating that the Commission's 2001 report recommended that:

> [i]t will be crucial for the public to be involved in the important decisions which need to be taken. We have to find a way to foster informed public discussion of the development and application of new technologies: whatever decisions are ultimately reached, they will be more palatable if they have not been taken behind closed doors.[71]

The AEBC report made explicit the recognition of the importance of public education and at least the feeling of engagement in decision-making of this type and it recommended active public participation in a national debate about GM crops.

Critical to that process is the question of risk perception. As described in the previous section of this chapter, there is a tendency to defer to 'experts', of which scientists and medical professionals certainly form a cohort. But for significant scientific breakthroughs in areas of intense ethical interest such as genetic modification or IVF, the issue becomes an intrinsically wicked one, where the scientific experts view the associated risks from an informed perspective not shared by the public. Often, the public perception is substantially informed by media coverage, which is why headlines such as 'GM food threatens the planet'[72] and 'M&S sells genetically modified Frankenplants'[73] tend to polarise opinion as much as inform it.

One analyst suggests:

> The technical definition of risk is 'the likelihood of adverse consequences from any hazard', but that is not the way the public sees risk. It does not explain why some risks trigger so much more alarm, anxiety or outrage than others, seemingly regardless of scientific estimates of their seriousness.[74]

The perception of risk in a simple yes or no decision is significantly complicated if there are diverse potential outcomes of which some are bad but very unlikely, and some are good and highly likely. More usually, there is some combination of good and bad outcomes with different degrees of likelihood. Expertise could be alleged to provide the basis for an objective decision amongst these competing outcomes:

> Many risk experts resist the pressure to consider [public] outrage in making risk management decisions; they insist that 'the data' alone, not the 'irrational' public, should determine policy. But we have two decades of data indicating that voluntariness, control, fairness, and the rest are important components of our society's definition of risk.[75]

Certainly that challenge was an element in the composition of the Warnock inquiry but, as Baroness Warnock insisted, 'it has to be recognized that there is no such thing as a moral expert'.[76] The deontological position seems to default to a demand for no risk of a morally unacceptable outcome (which will generally have zero probability), regardless of the potential benefits. Those who take this position seldom appear to carry responsibility for the provision of the potential benefits to society, and such dramatic oversimplification does seem to be characteristic of the majority of media coverage of such complex issues.

One could infer from the various (incomplete) records of the handling of the GM crops issue that government had long understood the importance of inclusive decision-making: '[p]ublic attitudes have long been identified as one of the key determinants of the development, application and subsequent technological evolution of technology'.[77] The AEBC report recommended the format that the GM debate should take and stipulated that it should 'include, but not be dominated by, the Government and current interest groups – the biotechnology and farming industries, NGOs [non-governmental organisations], and scientists. But to have public credibility and added value over the current level of debate, they must reach beyond these interests to a wider public.'[78] In the event, the AEBC commissioned a series of farm-scale evaluations in 2003–4 and simultaneously, the government commissioned a GM public debate as part of a wider GM dialogue which also incorporated a scientific review and a cost–benefit analysis; but was not made public. The public debate was held: '[o]ver a thousand people attended the six regional launch meetings and it has been estimated that a further 675 local meetings were organised across the UK'.[79] The steering board received 35,000 feedback forms and 1,200 letters or emails. A report was produced and, in 2004, after the farm-scale evaluations reported, the government simply aligned its position with EU legislation.[80]

What is pertinent here is the lesson that this exercise in public consultation provided for the government. In a specifically commissioned report, the Department for the Environment, Farming and Rural Affairs (DEFRA) noted 'the debate generated 'unprecedented levels of interest, participation and considered discussion about complex matters of science and policy amongst a relatively large number of the general public'.[81] That report acknowledged the context of wider government efforts to engage with the public on issues raised by scientific and technological development and, specifically, it comments that 'it is also important to consider what lessons can be learned from the debate process, to help inform future public engagement activities'.[82]

An independent commission was more credible in the eyes of the public, although more difficult to manage, and the process itself helped to educate the public. The minister responsible wrote to the chair of the AEBC in February 2001: 'I am convinced that the issue of separation distances is not simply a matter of science, but equally a question of public acceptability'.[83] Despite Mr Meacher's rather patronising view, the report recognised that the public had been more discriminating in its use of evidence than expected, noting '[i]t is clear that the public like to know who is advancing a particular view, and whether it is supported by evidence'.[84]

These episodes suggest increasingly sophisticated government understanding of the part played by pressure from the public in complex decision-making. The Warnock inquiry was a successful experiment in the use of 'experts' to inform both government and public in an ethically challenging area of technological progress. The GM debate specifically included the public as part of a wider dialogue within government and this appears to be a logical progression from the Warnock inquiry findings. It is also echoed by Lord Robertson's intent for the 1998 Strategic Defence Review (considered in chapter seven), for which he insisted on an extensive public engagement strategy.[85]

'Experts' and the public have different risk perceptions. The government reports quoted in this section suggest that the public is more discriminating when faced with conflicting evidence than government tends to give it credit for, but public understanding can be significantly affected by expert opinion and, it appears, media presentation. The handling of the GM debate hints that government tends to patronise the public and expect deference to experts if their opinion coincides with government inclination. The conclusion that appears to be drawn by several governments examined in this book is '[w]e take public concern very seriously, and we have weighed public opinion alongside the scientific evidence ... [we will] consider the best ways of providing the information which the public wants and in an open and transparent way'.[86] This conclusion was specific to the GM debate, but could reasonably be extrapolated to other wicked issues.

The nuclear debate

No similar inquiry has been established to consider UK nuclear deterrence. Arguably there is no need because there are only two departments of state to which nuclear deterrence is pertinent – the Foreign Office and the Ministry of Defence (coordinated through the Cabinet Office) – and there are processes in place which allow collaboration on policy formulation and the delivery of capabilities. Indeed, there have been numerous MOD and Cabinet Office inquiries into the strategic nuclear deterrent, but because so many of the details are classified, the findings are not made public, except in carefully edited inputs to public documents. One such was the 1980 open government document 80/23; but even this contained more information and detail than many of the Cabinet considered appropriate. The detail of how this was achieved and managed, and the associated studies and decisions prior to the deployment of Trident, were considered in chapter five, and more recent work will be considered in chapter seven. This is not to say that no inquiries or debates about nuclear deterrence have been held publicly by various organisations, merely that the government has not participated.

Notes

1. Michael Fallon, 'The case for the retention of the UK's independent nuclear deterrent: speech at the Policy Exchange', gov.uk (23 March 2016). Available: www.gov.uk/government/speeches/the-case-for-the-retention-of-the-uks-independent-nuclear-deterrent [accessed 22 November 2016].
2. King George VI, 'The King's speech to his peoples' (1939), London: BBC. Available: www.historic-uk.com/HistoryUK/HistoryofBritain/The-Kings-Speech/ [accessed 13 April 2013].
3. *Ibid.*
4. Alex J. Bellamy, 'The ethics of terror bombing: beyond supreme emergency', *Journal of Military Ethics*, 7 (2008), 41–65: 58.
5. *The Times*, 'Air bombing of civilians' (2 September 1939), p. 10.
6. Alex J. Bellamy, *Just wars: from Cicero to Iraq*. London: Blackwell, 2006, 5.
7. James Turner Johnson, *Ethics and use of force: just war in historical perspective*. London: Ashgate, 2011, 11.
8. Bellamy, *Just wars*, 5.
9. Quentin L. Quade, 'Civil disobedience and the state', *Worldview*, 10:11 (1967), 4–9: 5.
10. James F. Childress, 'Just war theories: the bases, interrelations, priorities and functions of their criteria', *Theological Studies*, 39 (1978), 427–45: 444.
11. S. S. S. Browne, 'Right acts and moral actions', *The Journal of Philosophy*, 42 (1945), 505–15: 509.
12. Michael Walzer, *Just and unjust wars: a moral argument with historical illustrations*. New York: Basic Books, 1977, 15.
13. *Ibid.*, 19.
14. Charles Guthrie and Michael Quinlan, *Just war. The just war tradition: ethics in modern warfare*. London: Bloomsbury, 2007, 23.
15. James Turner Johnson, 'Thinking historically about just war'. *Journal of Military Ethics*, 8 (2009), 246–59: 251.
16. Browne, 'Right acts and moral actions', 513.
17. David Fisher, *Morality and the bomb: an ethical assessment of nuclear deterrence*. Beckenham: Croom Helm, 1985, 25.
18. Jeff Mcmahan, *Killing in war*. Oxford: Oxford University Press, 2009, and David Rodin, 'Justifying harm', *Ethics*, 122 (2011), 74–110.
19. Johnson, 'Thinking historically about just war', 255.

20 I have not done this concept justice but would recommend the Appendix describing 'newspeak' in George Orwell's *Nineteen Eighty-Four* as an example of the use of language to circumscribe independent thought:

> The purpose of Newspeak was not only to provide a medium of expression for the world-view and mental habits proper to the devotees of Ingsoc, but to make all other modes of thought impossible. It was intended that when Newspeak had been adopted once and for all and Oldspeak forgotten ... a thought diverging from the principles of Ingsoc, should be literally unthinkable, at least so far as thought is dependent on words.

George Orwell, *Nineteen Eighty-Four*. Harmondsworth: Penguin, 1949, Appendix pp. 257–68.

21 See chapter 3 above; and Grayling, *Among the dead cities*; Slim, *Killing civilians*; Overy, *The bombing war*; Peter Gray, *The leadership, direction and legitimacy of the RAF bomber offensive from inception to 1945*. London: Bloomsbury Academic, 2012; and Peter Gray, 'The gloves will have to come off; a reappraisal of the legitimacy of the RAF bomber offensive against Germany', *Air Power Review*, 13 (2010), 9–40.

22 US National Conference of Catholic Bishops, 'The challenge of peace', para. 1.b.

23 Walzer, *Just and unjust wars*, 277.

24 Winston Churchill, *The gathering storm*. New York: Rosetta Stone, 2009, 492.

25 Walzer, *Just and unjust wars*, 247.

26 Michael Walzer, 'Political action: the problem of dirty hands', *Philosophy and Public Affairs*, 2 (1973), 160–80: 169.

27 Walzer, *Just and unjust wars*, 231.

28 Walzer, 'Political action', 171.

29 David Rodin, 'The moral inequality of soldiers: why *jus in bello* asymmetry is half right', in: David Rodin and Henry Shue (eds), *Just and unjust warriors: the moral and legal status of soldiers*. Oxford: Oxford University Press, 2008, 54.

30 Bellamy, 'The ethics of terror bombing', 42.

31 Walzer, *Just and unjust wars*, 262.

32 Anthony Kenny, *The logic of deterrence: a philosopher looks at the arguments for and against nuclear disarmament*. London: Waterstone & Co. Firethorne Press, 1985, 56.

33 Bruno Tertrais, 'In defense of deterrence: the relevance, morality and cost-effectiveness of nuclear weapons', Proliferation Papers (2011). Available: www.ifri.org/en/publications/enotes/proliferation-papers/

defense-deterrence-relevance-morality-and-cost [accessed 23 November 2016].
34 Hennessy, *The secret state*, 356.
35 Jonathan Haidt, *The righteous mind: why good people are divided by politics and religion*. London: Allen Lane, 2012, [Kindle edn] location 1671.
36 The ring of Gyges was a fable in which an imaginary magic ring made the wearer invisible; when invisible, the wearer ceased to behave morally. The point is: is it better to be seen as moral even though you act immorally, or to be seen as immoral even if you act morally? This goes to the heart of agency in ethics; is the actor the legitimate judge of morality, or is it for others to judge the actor? Plato, *The Republic*, Book Two.
37 C. West Churchman, 'Free for all', *Management Science*, 14 (1967), B-141–6: B-141.
38 Peter Riviere, 'Unscrambling parenthood: the Warnock Report', *Anthropology Today*, 1 (1985), 6.
39 E. O. Wilson, *The social conquest of Earth*. London: Norton, 2012, [Kindle edn] location 179.
40 Adam Briggle and Carl Mitcham, *Ethics and science: an introduction*. Cambridge: Cambridge University Press, 2012, 9.
41 Cabinet Office. 'Factors relating to the further consideration of the future of the United Kingdom deterrent', 7 December 1978. TNA DEFE 19/275.
42 Abse, HC Deb. 25 July 1978, *Hansard*, vol. 954, col. 669W.
43 Jackson, *ibid*.
44 MRC, Advisory Group to Review Policy on Research on *In Vitro* Fertilisation and Embryo Transfer in Humans: Minutes of the meeting held on Tuesday 6 March 1979. TNA FD7–2307.
45 Department of Education and Science, letter to MRC, '*In vitro* fertilisation: possible enquiry into medical ethics', 13 April 1982. TNA FD7–2307.
46 DHSS, press release 82/230 'Government enquiry into human fertilisation', 23 July 1982. TNA FD7–2307.
47 Fowler, HC Deb. 23 July 1982, *Hansard*, vol. 28, cols 328–30W.
48 Mary Warnock (chair). 'Report of the Committee of Inquiry into Human Fertilization and Embryology'. London: HMSO, 1984, para. 2.
49 *Ibid*., ch. 1 para. 7.
50 *Ibid*., 95–8.
51 Abse, HC Deb. 30 March 1982, *Hansard*, vol. 21, cols 279–86: col. 279.

52 Medical Research Council, letter D409/191, 22 November 1982. TNA FD7-2307.
53 Warnock Inquiry, letter, 'Research related to human fertilisation and embryology', 25 January 1983. TNA FD7/2307.
54 *Daily Mail* (18 October 1983).
55 Mary Warnock, 'Moral thinking and government policy: the Warnock Committee on Human Embryology', *The Milbank Memorial Fund Quarterly; Health and Society*, 63 (1985), 512.
56 *Ibid.*, 513.
57 Regis Olry, 'Body snatchers: the hidden side of the history of anatomy', *Journal of the International Society of Plastination*, 14 (1999), 6-9: 8.
58 *Lancet, The*, editorial: 'Traffic in dead bodies', *The Lancet* (1828), 818-21.
59 Olry, 'Body snatchers', 8.
60 Warnock, 'Moral thinking and government policy', 513.
61 *Ibid.*, 519.
62 *Ibid.*, 513.
63 Briggle and Mitcham, *Ethics and science*, 241.
64 Roger A. Pielke Jr, 'When scientists politicize science: making sense of controversy over *The skeptical environmentalist*', *Environmental Science & Policy*, 7 (2004), 405-17: 406.
65 Warnock, 'Moral thinking and government policy', 520.
66 *Ibid.*
67 *Daily Mail*, 'Meat may be tainted by Frankenstein food' (6 July 1999).
68 *News of the World*, 'Mutant porkies on the menu' (23 May 1999).
69 Food Standards Agency, 'Consumer views of GM food'. London: HMSO, 2003, para. 3.1.
70 AEBC, 'Crops on trial: a report by the AEBC'. London: HMSO, 2001, para. 2.
71 *Ibid.*, para. 68.
72 *Observer* (20 June 1999).
73 *Independent on Sunday* (18 July 1999).
74 Derek Burke, 'GM food and crops: what went wrong in the UK?' *EMBO Reports*, 5 (2004), 432-6.
75 Peter M. Sandman, 'Risk communication: facing public outrage', *EPA Journal (US Environmental Protection Agency)* (1987b), 21-2: 21.
76 Warnock, 'Moral thinking and government policy', 513.
77 Lynn J. Frewer, Chaya Howard and Richard Shepherd, 'Public concerns in the United Kingdom about general and specific applications of genetic engineering: risk, benefit, and ethics', *Science, Technology, & Human Values*, 22 (1997), 98-124: 98.

78 AEBC, 'Crops on trial', para. 69.
79 Department for Environment, Food and Rural Affairs (DEFRA), 'The GM public debate: lesson learned from the process'. London: HMSO, 2004, para. 2.
80 AEBC, 'Crops on trial'.
81 DEFRA, 'The GM public debate', para. 2.
82 *Ibid.*, para. 4.
83 Department for the Environment, Transport and the Regions, letter (Meacher, Minister of State DETR) to Chair AEBC (undated and unreferenced), repr. in AEBC, 'Crops on trial'.
84 DEFRA, 'The GM public debate', para. 22.
85 Lord Robertson, interview with Andrew Corbett, 12 November 2014.
86 DEFRA, 'The GM public debate', para. 1.

7

British nuclear deterrence in the 21st century

> We have to find a way to foster informed public discussion of the development and application of new technologies ... At present, there seem to be no avenues for a genuine, open, influential debate with inclusive procedures, which does not marginalise the reasonable scepticism and wide body of intelligent opinion outside specialist circles.[1]

This quote, from the 2001 government report on genetic modification of crops, seems to encapsulate a persistent problem for government that is prevalent in more than just the agriculture domain. The same frustrations appear perennially across Whitehall but there is no cross-government remedial action plan. In the absence of avenues for a genuinely open, influential debate on nuclear policy, it is striking that, as far as I can establish, no British prime minister has ever made a major speech on nuclear deterrence outside Parliamentary debate. Every French president since De Gaulle has made a keynote speech on 'their' nuclear deterrence policy in every term in office.

Is this important?

So far, this book has considered the factors that have affected British government engagement with the public on deterrence policy, and it has demonstrated that many appear to have their roots in the experience of the First and Second World Wars. Despite occasions where the government of the day has considered more active engagement with the public in order to explain or justify particular policy and procurement decisions, such engagement has seldom come to fruition; the most notable exception being the decision by Mrs Thatcher's government to publish OGD 80/23. Chapters two to five considered the historical evidence supporting this assertion.

In conclusion, this chapter will consider British government engagement with the public between the end of the Cold War and the 2021 Integrated Review, a period which has seen fundamental changes in the international security architecture and British defence, security and foreign policy objectives.

A 2015 YouGov poll conducted for *Prospect* magazine on the seventieth anniversary of the bombing of Hiroshima and Nagasaki in July 2015 was designed to gauge public attitudes to defence spending and the decision to replace Trident.[2] The survey suggests that about one-third of those polled thought the use of atomic weapons in the circumstances of 1945 was appropriate, about one-third thought it wrong and the other third had no definite opinion. Among those under 25, and amongst women, a higher proportion thought the use of atomic bombs was wrong. Only about one in ten think nuclear weapons make Britain less safe, with just under half thinking that they make us safer. About one-third would prefer like-for-like replacement for the Trident system, with just under one-third preferring a cheaper nuclear deterrent and about one in five advocating unilateral disarmament. In Scotland, about two-fifths advocate unilateral disarmament. Kellner concludes '[w]hether we regard nuclear weapons as a useful deterrent against bad behaviour or an insurance policy in case things go badly wrong, or simply as a symbol of national pride, most of us want the UK to keep them'.[3]

To a certain extent Kellner's terminology encapsulates one of the issues facing early 21st-century government; nuclear deterrence appears a very familiar and simple concept, but a dwindling minority of increasingly focused specialists is familiar with its key tenets and principles. Much of the commentary both inside and outside government is inconsistent and riddled with inaccurate language and misleading metaphors. Nuclear deterrence is not an insurance policy that mitigates outcomes if things go badly wrong; it is part of a range of mechanisms that help prevent things from going wrong. Caricaturing 'national pride' or 'prestige' can trivialise the genuine security and foreign policy benefits to the United Kingdom of a 'place at the top table'; being one of the five permanent members (P5) of the United Nations Security Council is not an inconsequential national interest.[4] There is, of course, no causative link between P5 membership and possession of nuclear weapons. Only the USA possessed nuclear weapons when it became a permanent member of the UNSC and those who suggest that continued possession is a factor in the UK's retaining its seat at that table have no basis for that assertion.

The use of inaccurate terminology is widespread: the options provided to its 'opinion-formers' panel' in a 2013 YouGov poll were: 'Britain should disband Trident and focus defence spending on conventional weapons and forces; Britain should replace Trident with a new and upgraded system; Britain should maintain Trident; Don't know; Other'.[5] The analysis of the poll published by YouGov stated: '[t]he most popular option, with thirty-six percent agreeing, was for Britain to "disband Trident and focus defence spending on conventional weapons and forces"'.[6] However, the poll actually showed that an absolute majority (52 per cent) of the opinion formers desired the retention of a nuclear deterrent (25 per cent felt it should be maintained and a further 27 per cent that it should be upgraded) but the inaccuracy of the language used by the poll skewed the result, and the analysis headline.

The choice that faced the government in 2016 was not to replace Trident – that system will continue until the 2040s when it will be upgraded – but whether to replace the submarines carrying it, which will reach the end of their service lives in the 2020s. The 2016 Parliamentary briefing paper 'Replacing the UK's Trident nuclear deterrent' perpetuated these inaccuracies, although it did state that '[a]lthough commonly referred to as "the renewal or replacement of Trident", the Successor programme is about the design, development and manufacture of a new class of four submarines'.[7] The executive summary for that briefing paper considers fourteen polls of public opinion since 2009 and suggests that

> a review of the available opinion poll evidence does suggest that, broadly speaking, the British public is divided on the question of whether Trident should be renewed. However, the public's views on

Trident are nuanced and their responses to public opinion polls are sensitive to the wording and framing of the question they are asked.[8]

Better understanding and information seem to be in order.

In a commentary on the result of the Brexit vote in the UK, populist Professor Brian Cox concluded that education is an important national security issue: '[w]hat you rely on in an open democracy is the ability of people to take an informed position but we're not teaching people how to think and we are becoming unstable as democratic societies'.[9] His core point was not that the state should teach the public 'what' to think, but 'how'; a democratic population must be equipped with the intellectual analytical tools to consider the provenance of assertions made in the public domain and determine which are valid, and which mere polemic. As the government report quoted at the head of this chapter notes, '[w]e have to find a way to foster informed public discussion'. This is particularly true for ethically complex policy areas, such as human embryology, genetic modification of crops and, of course, nuclear deterrence.

As I have mentioned several times, deterrence is a psychological process which influences the decision-making processes of a potential adversary. It does this by changing their risk analysis. The 2013 Trident Alternatives Review defined it:

> Deterrence rests on the notion of 'unacceptable loss' – the ability to inflict a level of damage that a potential aggressor would judge outweighed any benefit they might gain by a particular course of action.[10]

The more recent Defence Concepts and Doctrine Centre joint doctrine note 1/19; 'Deterrence; the defence contribution' resorted to the 1996 NATO definition: '[t]he convincing of a potential aggressor that the consequences of coercion or armed conflict would outweigh the potential gains'.[11] Deterrence is about the risk of the consequences, and the risks of damage to which an aggressor would be exposed through an attack against conventional defences are determined and controlled by the aggressor. It is subject to a straightforward risk–benefit analysis, and the aggressor can decide when to stop if aggression is proving too costly. If, however, the defender has nuclear weapons, the aggressor must consider the prospect that the defender reserves to themselves the ability to respond at a level of punitive retaliation entirely of their own choosing; the aggressor loses control of the level of risk to which they are exposed, and the risks are incalculably vast.[12]

This ownership of the level of risk to which the aggressor is exposed is the essential difference between policies of deterrence by denial (conventional deterrence; defence) and deterrence by punishment (the threat of retaliation). The process by which a potential aggressor makes decisions to coerce or act aggressively against a nuclear-armed state is hugely complicated by nuclear deterrence.[13] However, for most 21st-century states nuclear weapons have no unique military utility; there are very few military facilities that could not be successfully attacked with conventional means and the ability to destroy those facilities with nuclear weapons is no more of a deterrent than the ability to destroy them with precision-guided conventional munitions. The deterrent effect of nuclear weapons relies on something else. As Attlee grasped intuitively seventy-five years ago, 'the only deterrent is the possibility of the victim of such an attack [on cities] being able to retort on the victor'.[14] The difference in the political effect of nuclear deterrence is this unique fear that nuclear weapons engender. An aggressor must consider that as a last resort, and regardless of peacetime assurances, a nuclear-armed defender might target an aggressor's population centres. It is this visceral level to which Michael Fallon defaulted in his 2016 speech when he wondered 'how would the United States or France respond if we suddenly announced that we were abandoning our nuclear capabilities, yet will still expect them to pick up the tab and to put their cities at risk to protect us in a nuclear crisis?'[15]

Sir Michael Quinlan, the author of the 1980 Open Government document and doyen of the British nuclear strategy coterie for twenty years,[16] noted;

> the language was deliberately chosen – partly with ethical concerns in mind – to convey that while cities would not be guaranteed immunity, the UK approach to deterrent threat and operational planning in the Trident era would not rely on crude counter-city or counter-population concepts.[17]

Although not easily palatable, the cornerstone of UK national security strategy is nuclear deterrence, which is based on the contingent willingness to threaten civilian populations and, as Wyndham Goldie argued for *The War Game* (see chapter one): 'the people should be trusted with the truth'.[18] No modern democratic state openly advocates a deterrence policy that explicitly threatens non-combatants, and the examples from the First and Second World Wars in chapters two and three catalogue historical British reticence about exactly such thinking. But, as Fallon's speech shows, it is still – tacitly although probably not deliberately – confirmed in public statements. This does not mean that the strategy itself involves targeting population centres, but that an essential element of the psychological deterrent effect is that it might. In the rest of his 2016 speech, the most recent ministerial public speech on nuclear deterrence, Fallon went on to describe the British independent nuclear deterrent as 'minimum, credible, assured and independent'.

A minimum strategic nuclear deterrent

Much of the technical debate on Chevaline and the Polaris replacement (see chapters four and five) centred on the precept of a minimum deterrent. The 'Moscow criterion' was a measure of how little destructive power would be required to deter the most credible threat; at that time, the Soviet Union.[19] The 1980s debate was conducted in extreme secrecy; in 2007 Lord Owen argued that '[o]f course we had to keep it private. One was not in those days able to let the Soviet Union know that one had doubts about one's own deterrent capacity.'[20] He was making this argument in the context of his request for the government to release the papers from 1977–82, in order to inform a similar analysis of minimum deterrent for the 21st century, arguing '[a] minimum nuclear deterrent is not, however, a static concept. If we are to start, in 2010, the process of genuinely contributing to the elimination of nuclear weapons, it will not be credible if the British government commits to a new UK ballistic missile deterrent similar to Trident into the years 2050–60.'[21] Owen considered that the blanket application of 'secrecy' to the nuclear deterrent mission continues to be just as much an inhibition on effective public engagement as it was for Macmillan, Callaghan and Thatcher.

The 1998 Strategic Defence Review (SDR) announced '[w]e will retain our nuclear deterrent with fewer warheads to meet our twin challenges of minimum credible deterrence backed by a firm commitment to arms control'. The SDR reduced the UK stockpile from three hundred to fewer than two hundred operationally available warheads, and from a limit of ninety-six warheads to forty-eight per submarine, at a reduced day-to-day alert state.[22] The 2006 White Paper, which committed to the initial development of successor submarines to the Vanguard class, stated 'it will be the minimum necessary. We already have the smallest stockpile of nuclear warheads among the recognised nuclear weapon states and are the only state to have reduced to a single delivery system.'[23] Both the 2010 Strategic Defence and Security Review and the 2015 Strategic Defence and Security Review refer to a minimum deterrent,[24] without stipulating what level of destructive potential that involves, although both the 1998 SDR and the 2006 White Paper implied that the minimum deterrent required up to forty-eight warheads on each submarine,[25] and in 2010 this was further reduced to forty. Detailed study was conducted in support of these decisions, the nature and results of which are obviously extremely classified (as officer commanding an SSBN I was not privy to either the methodology or the details). The 2013 Trident Alternatives Review, conducted by Whitehall departments to meet the demands of the Coalition government, is a fascinating resource for research because it exposes more of the rationale, if not the numbers, behind such arcane calculations than do the White Papers: the key determining requirement for the Trident Alternatives Review was:

> [a] minimum nuclear deterrent capability that, during a crisis, is able to deliver at short notice a nuclear strike against a range of targets at an appropriate scale and with very high confidence. The study deliberately did not define 'minimum', 'short notice', 'scale' or 'very high confidence', as that could have overly constrained the list of system options for analysis.[26]

Intuitively, a minimum deterrent seems morally desirable over a more substantial capability, but a recent sophisticated analysis of contemporary US nuclear deterrence policy puts even this apparently simple logic under intense scrutiny:

> Minimum deterrence is a mode of deterrence that depends on the threat of nuclear retaliation alone and makes no effective accommodation for the principles of discrimination and proportionality or for a theory of deterrence that depends on putting at risk those things most valued by enemy leadership ... But retaliation against cities would violate the principles of discrimination and proportionality, and thus US deterrence strategy requires being able to put at risk other assets in an enemy country that enemy leaders would not want to lose.[27]

This seems to support the conclusion that minimum nuclear deterrence works because of the fear of an attack on cities. That said, American and British understandings differ on what causes wars and therefore how to manage and deter them, and this has a fundamental effect on their theories of deterrence and the associated ethics.

The USA provides 'extended nuclear deterrence' for a number of allies including NATO. For such extended deterrence to be credible to a potential adversary requires a capability to respond graduated in proportion to the aggression being deterred; a threat of a strategic nuclear response to a border incursion by a hostile reconnaissance aircraft into an ally's airspace would be literally incredible and morally untenable.[28] Thus, the ethical justification for 21st-century American nuclear deterrence presented by Roberts incorporates just war concepts of proportionality, and therefore discrimination, in a way that a British independent nuclear deterrence tailored as a last resort for national survival does not. In my experience, there is no British official synthesis of strategy and ethics at this level of sophistication.[29]

The 2021 Integrated Review is written in a very different style to its predecessors. Although continuing with the trend of expanding the analysis to include 'security' in the broadest sense and 'defence' and 'deterrence' as elements of a broader strategy to provide that security, it is couched very much in terms of 'power' and focuses on deterrence and defence almost exclusively in terms of military capabilities; it 'feels' more pugilistic than any review since the end of the

Cold War. The Prime Minister's foreword describes an integrated view of defence and deterrence: '[w]e will remain a nuclear-armed power with global reach and integrated military capabilities ... Our diplomacy will be underwritten by the credibility of our deterrent and our ability to project power.'[30] This apparent willingness to 'project power', with no real attempt to explain to what end, is disconcerting. In terms of delivering the nuclear elements of that deterrence posture, the review marks a radical shift from the trajectory of the previous thirty years.

In language almost identical to that used by NATO since the 1999 Alliance Strategic Concept[31] the review repeats previous defence review phrases: '[t]he fundamental purpose of our nuclear weapons is to preserve peace, prevent coercion and deter aggression'. It continues: '[a] minimum, credible, independent nuclear deterrent, assigned to the defence of NATO, remains essential in order to guarantee our security and that of our Allies'.[32] It does not describe what criteria are used to identify the minimum level of nuclear deterrence capability, but it does announce that the minimum will be increased, for the first time since the end of the Cold War. It asserts that some states' technologies, 'warfighting' nuclear systems and political rhetoric are changing (see 'A credible and assured strategic nuclear deterrent' below).[33] The review reiterates the declared strategy of continuous at-sea deterrence but states that:

> In 2010 the Government stated an intent to reduce our overall nuclear warhead stockpile ceiling from not more than 225 to not more than 180 by the mid-2020s. However, in recognition of the evolving security environment, including the developing range of technological and doctrinal threats, this is no longer possible, and the UK will move to an overall nuclear weapon stockpile of no more than 260 warheads ... We will maintain the capability required to impose costs on an adversary that would far outweigh the benefits they could hope to achieve should they threaten our, or our Allies', security ... While our resolve and capability to do so if necessary is beyond doubt, we will remain deliberately ambiguous about precisely when, how and at what scale we would contemplate the use of nuclear weapons. Given the changing security and technological environment, we will extend this long-standing policy of deliberate ambiguity and no longer give public figures for our operational stockpile, deployed

warhead or deployed missile numbers. This ambiguity complicates the calculations of potential aggressors, reduces the risk of deliberate nuclear use by those seeking a first-strike advantage, and contributes to strategic stability.[34]

The next section will consider in more detail continuous at-sea deterrence (CASD), which provides a strategy of assured response. Ambiguity is certainly a central feature of contemporary deterrence theory and it does complicate potential aggressors' risk calculations, but if that rationale is truly behind the announcement, why announce the increase in the stockpile? Why not simply announce the ambiguity and the decision not to be tied to previous ceilings? There appears to have been a rather simplistic link between number of warheads and deterrent value; the virility symbol much touted by CND. This numerical argument is demonstrable nonsense and was discarded in the 1980s but is unfortunately creeping back into deterrence discourse in the USA, and, it would appear, the UK. The ability to deliver an assured strike of sufficient destructive power is the fundamental capability underpinning minimum deterrence, but it requires a given capability against the benchmark adversary's defensive capability; not the adversary's offensive abilities, which is what the Integrated Review seems to suggest.

A credible and assured strategic nuclear deterrent

Uniquely amongst the nuclear weapon states, the UK has adopted a single nuclear weapon system: the Trident submarine-launched ballistic missile (SLBM). CASD is predicated on one submarine being perpetually at sea in a patrol posture and ready to fire; during the Cold War, that 'notice to fire' was measured in minutes; since the 1990s the missiles are not targeted and the readiness has been relaxed to days, not minutes. In a crisis, obviously, readiness could be reduced to minutes again with no external indications. The (classified) number of missiles and warheads embarked on each SSBN provides the minimum credible assured strategic nuclear deterrent; the stockpile is merely the capacity to manage those warheads.

The credibility of any deterrence policy depends on a combination of the resolve to carry it out and the capability; including readiness, reach, survivability, invulnerability and destructive power.[35] Reach, survivability and invulnerability are functions of the delivery system. Each system has strengths and weaknesses; ground-launched intercontinental ballistic missiles have a significant reach and in-flight survivability, but are vulnerable to attack before launch, unless fitted into mobile launchers; deployment of the launchers from their bases would be a very potent message in a crisis, perhaps seen as dangerous escalation, or as a demonstration of resolve; one of the paradoxes of deterrence is that the same action could indicate either or neither. Free-fall bombs and air-launched cruise missiles are as vulnerable as the aircraft on which they are embarked, and all aircraft are critically vulnerable on the ground; but both systems are flexible enough to deploy, or allow changes to readiness states or training patterns to be made obvious, in order to message resolve to an opponent in an extended deterrence strategy.

Vulnerability to pre-emption, or first strike, tends to demand very high-fidelity intelligence and warning capabilities, with a correspondingly high readiness of the nuclear forces, especially in a crisis. For this reason, a deterrence system based purely on vulnerable systems such as ground-launched missiles (especially those in silos) or air-launched systems, makes less contribution to strategic stability than an assured response system such as CASD. However, such systems also offer the capability to indicate resolve as an active part of a sophisticated deterrence-messaging system particularly appropriate for extended deterrence such as the USA provides for NATO.

A SLBM offers an assured response capability, and with a submarine always 'ready' providing CASD, it is invulnerable to a first strike. Contrary to the suggestion in the 2021 Integrated Review, while CASD contributes substantially to strategic stability – that is its major strategic attribute – announcing an increase in warhead stockpile numbers does not. CASD is not subtle and cannot be readily employed for messaging but it provides a credible threat of a guaranteed response *in extremis*, even if the parent state is attacked by surprise. The defender does not have to specify what this response would be, but a potential aggressor must consider the prospect that a nuclear weapon state faced with a threat to its vital national interests or its survival could retaliate against any element of the aggressor state.[36]

The perception of the agent being deterred is critical; if the defender's resolve is not evident to the aggressor before the aggressor makes decisions, a policy of deterrence is much less likely to be effective. The everyday peacetime communication of capability and resolve is therefore fundamental to a policy of deterrence. Contemporary British defence policy is more predicated on deterrence than ever before, with the Foreword to the 2015 Strategic Defence and Security Review (SDSR) stating:

> we have reconfigured Britain's Armed Forces so they are able to deal with modern and evolving threats. Where necessary, we will be ready to use force ... we will use ... our soft power to promote British values and to tackle the causes of the security threats we face, not just their consequences.[37] [The theme of deterrence in preference to conflict was repeated throughout SDSR 2015.] We will strengthen our Armed Forces and our security and intelligence agencies so that they remain world-leading. They project our power globally, and will fight and work alongside our close allies ... to deter or defeat our adversaries.[38]

SDSR 2015 introduced an interpretation of deterrence that appears to move away from 'conventional deterrence' and 'nuclear deterrence' to deterrence using the 'full spectrum of our capabilities to ... deter potential adversaries, including through renewal of our nuclear deterrent'.[39] This was depicted within a 'fusion doctrine' in the 2018 National Security and Capabilities Review which further enthused about this 'whole of government' approach, without explaining how it would deter.[40]

This contemporary interpretation of the concept of deterrence stems from a project led by the Defence Science and Technology Laboratory and potentially heralds one of the most fundamental changes in national deterrence and defence policy in recent years. It is a significant evolution and will require more than a simple statement to enable it; coherent 'full spectrum' deterrence will require departments of state to coordinate their activities in order to achieve a tailored effect against a potential adversary that some may not yet consider a problem. One early step towards this has been the formation of the Cabinet Office Communications Steering Group which includes representation from Cabinet Office, MOD, FCO and others as required.

This group was formed in 2015 after an increasing recognition in the nuclear policy sections of the MOD and FCO that the departments had effectively been deterred from engaging with the public at all on nuclear deterrence matters since the Civil Defence Handbook debacle of 1980. The imperative for this change to the status quo was the imminent decision to commit to the next stage of the development of a successor submarine to the Vanguard class (the Dreadnought class). This recognition was compounded by an acknowledgement that there was a corresponding lack of understanding of many aspects of British nuclear deterrence policy and programme not only among the public at large, but throughout Whitehall and Westminster; at all levels. The 2019 public commemoration of the fiftieth anniversary of the CASD operation (the UK's longest military operation) was a low key experiment in official endorsement, if not exactly celebration, of the mission and culminated in a service of thanksgiving at Westminster Abbey.

In keeping with its more pugilistic tone, the 2021 Integrated Review states:

> We must update our deterrence posture to respond to the growth in state competition below the threshold of war under international law. As set out in the 2020 Integrated Operating Concept, this means being able to move seamlessly between 'operating' and 'warfighting'. This will require a force structure that principally deters through 'persistent engagement' below the threshold of war, while remaining prepared for warfighting when necessary.[41]

It goes on to assert that

> [s]ome states are now significantly increasing and diversifying their nuclear arsenals. They are investing in novel nuclear technologies and developing new 'warfighting' nuclear systems which they are integrating into their military strategies and doctrines and into their political rhetoric to seek to coerce others. The increase in global competition, challenges to the international order, and proliferation of potentially disruptive technologies all pose a threat to strategic stability.[42]

The supporting Command Paper 'Defence in a competitive age' does not really expand on this, nor does it yet describe a coherent deterrence strategy based on a seamless transition between operating and warfighting, least of all how that relates to an assured response nuclear deterrent. It does emphasise the international collaborations involved in the UK strategic nuclear deterrent:

> We will also replace the UK's nuclear warheads to ensure we maintain an effective deterrent throughout the commission of the Dreadnought Class, working closely with the US so that our new sovereign warhead remains compatible with the Mk7 aeroshell and Trident Strategic Weapon System. We will continue to cooperate with France under the Teutates Treaty, working together on the technology associated with nuclear stockpile stewardship in support of our respective independent nuclear deterrent capabilities.[43]

An independent strategic nuclear deterrent

Since the Labour party argued in 1964 that '[Polaris] will not be independent and it will not be British and it will not deter',[44] there has been a perception that the British nuclear deterrent lacks independence. Greenpeace argue 'Trident is not a "UK" weapons system – this is another well-worn myth. Nearly all of the weapon parts are sourced or leased from the US, with few exceptions, such as the warhead, is manufactured in the UK based on the US W76 design. The software, targeting and weather data are all US-sourced.'[45] CND suggest that '[o]ur nuclear weapon system is neither politically or technically independent. It has been assigned to NATO since the 1960s, meaning Trident could be used against a country attacking – or threatening to attack – one of the alliance's member states.'[46]

This is true; the UK does participate in the collective defence of NATO allies, and British forces form part of the NATO deterrence and defence posture; this has been the basis of British defence policy since 1949. Some British forces are assigned to NATO and come under NATO command; the UK SSBNs, however, are not assigned to NATO. The implication that there is a loss of control over the decision to launch a strike is misleading; decision-making is a fully sovereign UK function. Only the British Prime Minister or their nuclear deputy can authorise the launch of the UK's nuclear deterrent missiles; NATO can only request it.[47] CND also assert that the system is 'dependent on US technical support …The British submarines must also regularly visit Kings Bay [the US Trident operating facility in Georgia, USA] for the maintenance and replacement of these missiles.'[48]

Commons Defence Committee evidence suggests otherwise: 'operationally the system is completely independent of the United States. Any decision to launch missiles is a sovereign decision taken by the UK and does not involve anybody else.'[49] The British reliance on the USA for deep technical support for missiles is true, in the sense that the missiles receive maintenance only in the USA, not while they are embarked on British submarines. However, they remain embarked for the duration of each operational commission, between twelve and fourteen years.[50] In every operational sense, UK Trident is fully independent. Although official references to the independent nuclear deterrent are almost ubiquitous, dependency is an example of a myth that, once created, seems to be difficult to refute. My own experience is that the UK and the USA train on and maintain the same systems to the same standards; but operate completely independently. The 2010 Teutates agreement provides for Anglo-French collaborative research facilities into warhead development, with the two parties sharing updated test and development laboratory facilities rather than duplicating facilities and expense.[51]

A reluctant nuclear weapons state

From the very start of its nuclear weapons programme, Britain has been a reluctant nuclear weapon state. Attlee, even as he defined the essence of deterrence, sought a means of inhibiting the use of war as a tool of policy in future, suggesting that '[a]ll nations must give up their dreams of realising some historic expansion at the expense of their neighbours'.[52] Only after strenuously seeking some form of international agreement did his government recognise that the UK 'should itself undertake the production of atomic bombs as a means of self-defence as soon as possible'.[53] This 'twin' focus on some form of international control of nuclear weapons has been an element of British thinking about nuclear weapons ever since, although it has been peripheral to this consideration of the historical development of nuclear deterrence policy.

Each of the Defence Reviews since the end of the Cold War has addressed nuclear deterrence in the context of a commitment 'to working towards a safer world in which there is no requirement for nuclear weapons' and made corresponding assertions about disarmament achievements.[54] These are closely linked to perceptions of minimum deterrence, maintaining the lowest force levels necessary to meet that capability. After the 2006 White Paper which commissioned the early development of the Dreadnought class, the FCO published 'Lifting the nuclear shadow', a concise appraisal of the issues surrounding nuclear arms control, cataloguing the aspirations for a nuclear weapons-free world. In the foreword, then Foreign Secretary David Miliband wrote:

> These are issues which do not just concern Foreign or Defence policy. They are about the security of our world both now and in the next generation and deserve wider engagement. I therefore asked for the issues to be set out in a way that does not expect the reader to know the subject inside out ... The path to eliminating all nuclear weapons. To achieve this will require bold thinking and careful work by many nations. The UK is wholeheartedly committed to playing its part in this process.[55]

'Lifting the nuclear shadow' included an explanation of international agreements and Britain's role in them, a consideration of the concept of 'Global Zero' (a reduction to no nuclear weapons worldwide) and a consideration of the implications of security without nuclear weapons. It was soon followed by a government-wide document 'The road to 2010',[56] which set out the British position in advance of the 2010 Nuclear Non-Proliferation Treaty conference. Unfortunately, this document was less clear and much of the elegant alignment of deterrence policy and disarmament aspiration described in 'Lifting the nuclear shadow' was muddled and the message confused or lost. It is this type of incoherent official contribution to public discourse on nuclear deterrence that the Cabinet Office Communications Steering Group should avert.

Ten years later, the 2020 Nuclear Non-Proliferation Treaty conference was postponed due to the coronavirus pandemic and rescheduled for August 2021, six months after the publication of the 2021 Integrated Review. There is no mention of the NPT conference in the Integrated Review, though the supporting 'Defence in a competitive age' paper states that '[w]e remain committed to the long-term goal of a world without nuclear weapons and continue to work for the preservation and strengthening of effective arms control, disarmament, and non-proliferation measures, taking into account the prevailing security environment'.[57] There is no indication that this is anything more than a formulaic platitude, and there is no indication that any serious effort is being expended to identify the contours of a 'world without nuclear weapons'.

Factors in the development of British nuclear deterrence policy

Strategic imperatives

The rationale that the 'bomber would always get through' was born in the perceived inability to respond effectively to the Zeppelin raids of 1915 and the Gotha raids of 1917. It is the essence of the deterrence argument; Baldwin's famous dictum continued '[t]he only defence is offence. You have to kill more women and children more quickly than the enemy if you want to save yourselves. I just mention that … so that people may realize what is waiting for them when the next war comes'.[58] This was borne out by the reports from Guernica and China. In the First World War, the Cabinet had been dealing with a wicked problem; technological change had enabled warfare to evolve to challenge accepted moral standards, and there was a clear demand from an increasingly vociferous lobby to retaliate in kind, and a strategic imperative to combat such tactics. A lack of response might damage already strained morale among the British population; defence against such attacks was limited, but to retaliate was repugnant to many in office. In the event, limited retaliation was conducted, but in the context of extreme secrecy and an almost apologetic government narrative.

In 1940, Churchill's Cabinet was faced with a similar dilemma; after Dunkirk the only means of offensive action against Germany was to bomb its industrial capabilities, but although the RAF had actively prepared for just such a campaign between 1928 and 1940, the British bombing forces were technologically incapable of mounting a sustained bombing offensive, and the scant resources were in demand for every other task as well. By 1942 'Bomber Command provides our only offensive action yet pressed home directly against Germany. All our other efforts are defensive in their nature, and are not intended to do more, and never can do more, than enable us to exist in the face of the enemy.'[59] There is no evidence that suggests that had Britain had the capability to conduct precision bombing and avoid collateral casualties in cities, it would not have done so, although some advisers did advocate this: '[i]f even half the total load of 10,000 bombers were dropped on the built-up areas of these fifty-eight German towns the great majority of their inhabitants (about one-third of the German population) would be turned out of house and home'.[60] The gradual build-up of Bomber Command between 1941 and 1943 enabled the vision of a substantial bombing campaign against German industry, but although the new heavy bombers were capable of delivering substantial payloads, they were very vulnerable to air defences during daylight. Their navigation was still not accurate enough to conduct night attacks and achieve the precision bombing which the British government had used as the cornerstone of the moral case for the offensive. Britain therefore made the decision to shift to night bombing, accepting the reduced accuracy in order to reduce RAF casualty rates.

The evolution of NATO nuclear strategy during the Cold War was almost entirely driven by the USA. That said, Britain played an important role, particularly during the 1960s, with Healey advocating the graduated deterrence strategy which evolved into NATO's flexible response. During the late 1960s and early 1970s, this further developed into a nuclear warfighting strategy (using nuclear weapons for military effect, not just deterrence) which survived until the INF treaty in 1987 and the demise of the Soviet Union. Since the Cold War, NATO has adopted an explicit policy of nuclear deterrence; the 1991 Strategic Concept stated: 'Allies concerned agreed to move away, where appropriate, from the concept

British nuclear deterrence in the 21st century 195

of forward defence towards a reduced forward presence, and to modify the principle of flexible response to reflect a reduced reliance on nuclear weapons'.[61] This suggests that NATO still envisaged the use of nuclear weapons for military effect, albeit with reduced reliance on them.

In 2010 this evolved into '[d]eterrence, based on an appropriate mix of nuclear and conventional capabilities, remains a core element of our overall strategy'.[62] While this may appear semantic, the change in terminology, with deterrence elevated above defence, represented a profound change for NATO. The Warsaw Summit of 2016 saw intense negotiation on the 'nuclear language' to be included, evolving to a more robust, but non-confrontational, statement of resolve:[63] '[t]he fundamental purpose of NATO's nuclear capability is to preserve peace, prevent coercion, and deter aggression. Nuclear weapons are unique. Any employment of nuclear weapons against NATO would fundamentally alter the nature of a conflict.'[64] This text implied that alliance nuclear weapons remain for deterrence purposes, and the 2018 summit communique trod an equally fine line between resolve and restraint.

The 2021 Integrated Review suggested that

> UK policy has been focused on preserving the post-Cold War 'rules-based international system' which has greatly benefited the UK and other nations. Today, however, the international order is more fragmented, characterised by intensifying competition between states over interests, norms and values. A defence of the status quo is no longer sufficient for the decade ahead ... In particular, we will increase our efforts to protect open societies and democratic values where they are being undermined.[65]

This is an important contextual observation for the subsequent announcements about the increases in nuclear weapon stockpiles. It suggests that the UK perceives an enhanced sense of competition in the world of 2021 and is preparing to take a more belligerent attitude within it. As the cornerstone of a national deterrence posture, this is critical, if it is perceived as credible by potential adversaries. Such credibility will require genuinely coherent strategic action, and not the selectively reactive interventionism of the last twenty years. In combination with the announcements about the nuclear weapon stockpile, this suggests a clear break from the restrained resolve of the NATO-wide response to events of the last decade. It is not a clear indication of a commitment to strategic stability.

Industrial capacity

This research has found no historical evidence that suggests that industrial or technical capacity in itself is a factor in the decision to become or remain a nuclear power. Arguably this might be related to 'prestige' which is considered in the next-but-one subsection. Advances in high-technology industrial capacity are acknowledged as a valuable side-product of nuclear-related work, but not the motive for it. In the Trident decision of 1980 and the recent successor debate, industrial capacity and the limited facilities to construct nuclear submarines have been noted as limitations on decision-making and timelines: 'greater industrial collaboration and affordability are essential components in any new submarine programme and … it needs to address its own shortage of skills in managing a programme of the scale of a Vanguard successor'.[66] The BASIC Trident Commission concluded '[i]ndustrial and economic impacts are important for the communities concerned, but cannot play a key role in determining whether the UK continues to deploy a nuclear deterrent'.[67]

Cost – a technical factor?

In the early days of the atomic programme, British national strategy was simple – atomic bombs dropped from heavy bombers would act as a deterrent. Budget restrictions meant, however, that the bomb technology would not be paired with a suitable bomber until 1957, by which time the atomic bomb was perceived as obsolete in the face of the hydrogen bomb, and the heavy bomber–bomb combination was obsolescent in the face of developing missile technology. The pace of technological change and perceived associated strategic imperatives drove the UK into the position where Britain could no longer afford to develop a sovereign nuclear capability;

British nuclear deterrence in the 21st century 197

and as described in chapter five, purchased the Polaris system from the USA.

Similar combinations of budget and technical strategic imperatives have dominated British national system choices ever since, in particular the decision to purchase Trident. Trident D5 was acknowledged to be more precise and capable than the minimum deterrent required by Britain, but in terms of cost and 'future-proofing' it would have been ludicrous to reject it in favour of a less capable system that would have been more expensive. In one sense, a similar situation pertained in 2016. Simply as a technical planning factor, cost suggested that the cheapest option available to sustain the current level of deterrence capability was to replace the submarines carrying Trident. The supporting work for the 2006 White Paper included a very substantial study which considered over 100 different combinations of theoretical and existing weapons and delivery systems. This study reviewed technical risk, cost and effectiveness, including scale (minimum destructive capability), independence, range, vulnerability and readiness. These options were narrowed down to four generic options: long-range aircraft equipped with cruise missiles; surface ships equipped with Trident missiles; a land-based (silo) system equipped with Trident missiles; and submarines equipped with Trident missiles.[68] The political decision to proceed with what became the Dreadnought project was the result of that choice.

Cost can also, obviously, be used as a tool of political argument. Much of the anti-nuclear opposition of the 1950s and 1960s was to the possession of nuclear weapons in the first place; a moral position. In the 1980 decision on Trident, this remained an element but the nature of the opposition was coloured by the NATO LRTNF decision of December 1979, the siting of cruise missiles in the UK and an increased public perception of the immanence of the threat of nuclear war. The moral argument was substantially displaced by fear of nuclear war and the lack of informed official intervention contributed to the conflation of nuclear war with nuclear deterrence that has dominated the public discourse ever since.

There was opposition to the decision to replace Polaris with Trident on cost grounds, but it was not a pre-eminent aspect of the anti-nuclear platform. Since 2006, however, there has been sustained and coherent criticism of the Dreadnought decision on cost grounds. For example, CND maintains that the costs will be 'at least £205 billion. This money would be enough to improve the NHS by building 120 state of the art hospitals and employing 150,000 new nurses.'[69] CND does not make clear that the headline figure is calculated taking into account the entire capital, through-life running and disposal costs into the 2060s.[70] The government publishes costs in terms of the way it budgets which, in this case, is for the capital cost; 'as set in the 2015 SDSR we estimate that 4 new Dreadnought submarines will cost £31 billion to build, test and commission, spread over 35 years, with a contingency of £10 billion. On average, that amounts to 0.2% per year of government spending.'[71] Officials and government politicians tend not to share platforms with CND lobbyists, so these figures are seldom tested like for like.

However, cost should become a factor only if the principle of possession of nuclear weapons is accepted. If it is, then a cost–benefit analysis of various systems is entirely pertinent and an appropriate function of Parliamentary debate and oversight. It is notable that the focus of anti-nuclear opposition has moved away from the principled opposition to nuclear weapons of the 1958 CND policy statement,[72] which has had limited public resonance despite the 1960 and 1980 peaks, to arguments about costs. CND also asserts that '[f]ormer Prime Minister Tony Blair, one of the biggest supporters of replacing Trident in 2007, has admitted that the only purpose of maintaining the nuclear weapons system is to give Britain status'.[73]

Prestige

One of the most invidious criticisms of British nuclear deterrence is that it is a virility symbol; a sop to politicians' vanity. Healey said of Macmillan's government 'they clutched at the nuclear missile as a virility symbol to compensate for the exposure of their military impotence at Suez'[74] and Ruddock asserted in Parliament that 'the Government are committed to a massively expensive useless virility symbol called Trident'.[75] The virility critique is a corruption of the suggestion that nuclear weapons give a state a unique status, or prestige. This is a perpetual, but discreetly acknowledged factor. Macmillan briefed his Cabinet in July 1954: 'unless we possessed thermo-nuclear weapons, we should lose our influence and standing in world affairs'.[76] Ironically, Healey – once Defence Secretary – noted 'a few Polaris submarines would be worth more than the same number of hunter-killers ... because they would give Britain more influence, particularly in Washington'.[77] Alec Douglas Home said, when interviewed by Robin Day, 'I do believe that if we deprive ourselves of all control over our nuclear arm then Britain becomes a second-class Power'.[78] This different interpretation of virility was prevalent at all levels; the Chiefs of Staff advised Macmillan's Defence Policy Committee '[t]o that end we must strengthen our position and influence as a world Power and maintain and consolidate our alliance with the United States'.[79] One of the unanswered questions to Mrs Thatcher's Cabinet raised by Armstrong was: '[g]iven the decline in our world position in other respects, will it do us enough good to stay in the league from the 1990s to justify the cost of the burden?'[80]

The situation pertains today. Lord Robertson recalled the Shadow Chancellor Ed Balls arguing at the Trident alternatives review:

> it is nothing to do with defence; at the end of the day it is the argument that nobody can use in public, that if you opt out of the nuclear club then you opt out of senior rank in the World. It's the price tag; otherwise the French have it, the only European nuclear power, the Americans will feel betrayed and therefore you just relegate us.[81]

Although there is no evidence that nuclear weapon status is related to permanent UNSC membership, *The Times* concluded in 2009 that

> Trident and its successor are as much about national power and Britain's position in the world as about military effect. The five permanent members of the UN Security Council (the US, Russia, China, Britain and France) achieved their positions by being the victors of the Second World War. But they now retain those seats only thanks to their possession of credible nuclear deterrents ... abandon the deterrent and, sooner or later, Britain loses its seat.[82]

The issue of prestige is not simple, nor is it superficial; Fox argued 'a global role for the UK is a necessity, not a luxury ... Britain must help shape a changing world, rather than merely react to it'.[83] The 2010 SDSR stated:

The National Security Strategy sets out two clear objectives:

(i) to ensure a secure and resilient UK by protecting our people, economy, infrastructure, territory and ways of life from all major risks that can affect us directly; and

(ii) to shape a stable world, by acting to reduce the likelihood of risks affecting the UK or our interests overseas, and applying our instruments of power and influence to shape the global environment and tackle potential risks at source.[84]

This is sustained in SDSR 2015 where Britain '... plays a strong, positive global role. We project our power, influence and values to help shape a secure, prosperous future for the UK and to build wider security, stability and prosperity. We have unique strengths that enable us to do this.'[85] If Britain wants to shape a changing world, rather than merely react to the changes, then a seat at the top table, or prestige, is an invaluable and irreplaceable asset in the national 'toolbox'. Many senior politicians, including some who previously ridiculed the idea, appear to believe that 'prestige' is related to the nuclear deterrent and merits consideration in its retention.

Civil defence

Civil defence has not really been considered as a genuine factor in nuclear deterrence strategy since the debacle of the early 1980s. The derision to which the publication of 'Protect and survive' exposed the concept was sufficient to ensure the end of the civil defence project. To a certain extent, this process had been ongoing since the mid-1950s when the Strath Report highlighted the inability to provide any credible defence against nuclear weapons, except for a select few in deep underground facilities. The choice between preparation to fight a nuclear war and investment in deterring it was really made in 1955 and tacitly made public in the Sandys Defence Review of 1957. *The War Game* was an unwitting step in the process, ridiculing the Civil Defence Organisation claims that nuclear war would be survivable. It stemmed from the recognition that deterrence was more credible than defence, but that this would be difficult to present in public, as the Cabinet Office recorded in 1954:

> These and other changes recommended in this report certainly could not be defended in isolation. Public acceptance of them can only be secured if they are presented as parts of a coherent plan based on the recognition that no purely defensive policy could ensure the safety of these islands and those who live in them and that the main weight of our defence effort must now be concentrated on building up the deterrent strength which will prevent the outbreak of a major war.[86]

Despite officials' advice, the contents of the Strath Report and their ramifications remained closely guarded secrets in Whitehall in order to avoid the need for awkward explanations of meaningless shelter and evacuation policies.[87] The publication of the Civil Defence handbook by the Home Office in 1980 was an own goal of epic proportions. Survival in a nuclear war provided plot lines for much of the fiction produced at the time and civil defence was a staple for comedy such as *Yes, Prime Minister*. The irony is that when given the opportunity to engage in informed debate after seeing *The Day After*, American audiences proved to be rational and well informed, and not radicalised to one extreme or another. This suggests that the public should be trusted with the truth and enabled by government to make an informed opinion; this is not yet happening in the UK.

Moral views

The tension between the ability of modern warfare to bypass traditional concepts of battlefields, combatants and non-combatants, and the ethical implications of doing so, has been at the centre of British strategic thinking since 1915, both in public and in Parliament. 'The invention of the bombing plane abolished chivalry for ever. It is now "retaliate or go under".'[88] In 1915–18, there was no evidence that it was 'retaliate or go under', but there was a determined minority that advocated retaliation for what was commonly perceived to be the 'haphazard murder' of aerial bombardment by German aircraft.[89] A similarly vociferous lobby advocated a strongly deontological (some things are just wrong) position: 'does the Government think that, if we send aeroplanes to kill little innocent German babies, that is going to help the situation?'[90] Neither side really engaged much with the other, and the government, which was trying to sustain a war the like of which had never been seen before, was caught between two mutually exclusive positions.

During the interwar years, although states unsuccessfully sought to outlaw aerial bombardment, the embryonic Royal Air Force adopted the doctrine as its *raison d'être*. But this doctrinal focus was not successfully converted into capability. The debate at the COS meeting in 1928 is instructive for the continuity of issues associated with strategic bombing and (ultimately) nuclear deterrence. Trenchard argued that while it was immoral to bomb cities purely in order to terrorise the civilian population, bombing which interrupted manufacture, transportation etc., which also terrorised the civilian population, was legitimate.[91] During the Second World War this was the strategy ostensibly followed by the RAF, although casualty rates during daylight raids forced a shift to less accurate night bombing. The issue here, of course, was the balance between casualty rates among aircrew and the desire to minimise civilian casualties. During the same period the US Army Air Force, faced with similar factors, chose to continue with daytime bombing, and developed additional protections for the bombers.

There were serious tensions between the leadership of Bomber Command, the senior leadership of the RAF and the Cabinet over the public presentation of the activities of Bomber Command: '[a]ny public protest, whether reasonable or unreasonable, against the bomber offensive could not but hamper the Government in the execution of this policy and might affect the morale of the aircrews themselves'.[92] Modern analysts of the strategic bombing campaign tend to describe the moral thinking that enabled it as the 'supreme emergency' idea: 'what we fight for is of such ultimate importance that we have to break some of our own rules to defend it'.[93] This does not sit well with modern rights-based ethics which admit of no such violation.

Although there may be elements of the classified nuclear briefing provided to senior ministers which deals with these questions, I have found no evidence of rigorous ethical consideration of the nuclear deterrence mission, although Macmillan's account of Churchill brooding a good deal about the atomic and hydrogen bomb does support anecdotal evidence that individuals did and do genuinely agonise over these issues.[94] My experience was that nothing was provided to SSBN commanding officers between 2003 and 2009. The supreme emergency logic is one to which I believe many involved in the nuclear deterrent mission would subscribe, supported by the less well-articulated feeling that deterrence of war by those with 'dirty hands' imposes peace, which is a desirable moral good in itself.

The theologian Robert Fitch, writing in 1940, supposed that '[f]or if history means anything, then Jesus developed his thought by criticizing traditional materials in the light of the challenge of new circumstances. This is apparent everywhere.'[95] This argument to adapt traditional moral thinking to prevailing circumstances seems to undermine the basis and character of any kind of moral certainty, and reflects Johnson's view on the mutability of the just war tradition. But, Fitch continued: '[d]oubtless, in a world of fixed structure and pattern we should find certainty in a body of fixed moral principles; but in a world that is characterized by growth, plasticity, and emergence such a set of principles can lead only to chaos and confusion'.[96] The tendency of modern ethical study to seek fixed normative prescriptions in a rapidly evolving social and international environment seems to meet Fitch's view of principles which are largely illusory and of limited utility to practitioners.

Much more useful, even to a non-Christian, is Butterfield's 1951 analysis of St Augustine: 'love God and do what you like'.[97] This would not support pithy public debate – a moral framework of this sort is not conducive to glib simplification – but it would enable detailed consideration of the factors in play at the moment of decision in highly complex situations without having to reduce this to public soundbites such as 'does the Government think that, if we send aeroplanes to kill little innocent German babies, that is going to help the situation?'[98] Such a framework could be construed as a miscreants' charter, but if the essential morality of the decision-makers is accepted, and their decisions are tested rigorously in specialist environments, it provides a better test for practical ethics than the current arrangement where engagement is eschewed altogether.[99] Ethicists and practitioners need never agree, but the conversation is critical and, at present, seems to be missing.

Experts and the public debate

The Warnock inquiry was a successful experiment in the use of experts to inform both government and public in an ethically challenging area of technological progress. This inquiry model is unlikely to be successful in the public analysis of defence policy because of the security considerations; even if the experts were exposed to the classified material, their report could be no more classified than official public engagement. Recent examples of official inquiries could include the Chilcot inquiry into the 2003 Gulf War which was commissioned in 2009 and reported in 2016, and the Saville inquiry into the 1972 'Bloody Sunday' which was commissioned in 1998 and reported in 2010. Such timescales are not conducive to policy formulation, and information that would remain classified during and after the inquiry could not be included (even redacted) in the public reports.

Nott's reluctance to include a second minister in the Trident debate in 1981 is striking; he did not believe that two ministers could 'master the brief'.[100] Mrs Thatcher did not agree. Given that the core (if not the whole) of British defence policy in 1980 was maintenance of the security of the NATO alliance through deterrence of the USSR, it is surprising to note the degree of ignorance of the key aspects of the deterrent which Nott seemed prepared to countenance, among the very ministers expected to oversee it. The current provision of nuclear deterrence training for military officials is negligible. There is one half-day session allocated to the (year-long) Defence Academy 'Advanced Command and Staff Course' (ACSC) and no specific policy training provided for those involved in the mission.[101] Officials joining departments of state which deal with the nuclear deterrent are expected to pick up the details in post. In contrast, CND offers a series of bespoke educational packages for schools aligned with the National Curriculum for various subjects including 'Drama, English, Citizenship, Religious Education, Maths, ICT, Computer Science, Art & Design, Government and Politics, Social Sciences, History'.[102] Such educational facilities are not available from official sources.

A further challenge is one highlighted by Warnock; there is no such thing as a moral expert.[103] While the public may defer to doctors, engineers or airline pilots as experts, whose expertise is technical, there appears to be a much less deferential attitude to moral issues where the acknowledged technical expertise challenges intuitively held moral convictions. This is even more pronounced in the rather febrile social media environment of the UK in 2021 and the populist challenge to the opinion of experts epitomised by the Brexit campaign. Informed official engagement on complex moral issues is more important than ever.

> Insofar as possible, it is wise to simplify language rather than content – that is, take the extra words to make hard ideas clear. Unfortunately, neither the expert source nor the lay audience is usually willing to dedicate the time needed to convey complex information a step at a time.[104]

Sandman was writing about public perceptions of risk, but the principle holds for emerging technologies, such as nuclear weapons. The onus is on the experts to make their expertise accessible. At present, they struggle and the situation is exacerbated because the majority of the public, and decision-makers, receive information filtered through mass media, or skewed by social media. The importance of critical engagement with media is difficult to overstate.

The 1998 Strategic Defence Review deliberately drew on extensive public consultation. The review was programmed to take twelve months; in the event it took eighteen. Lord Robertson subsequently noted:

> we would invite everyone, the CNDers, the unions, journalists, academics ... we wanted to be as open and inclusive as possible, both because it was a good thing to do and partly because I thought

'I want people at the end of the day to say it's our review' and I joke now and say that I had always said that if the SDR was successful, it would be called the SDR and if it was a failure it would be called the Robertson review.[105]

This inclusive engagement was reflected in the Commons Defence Committee which had eleven sessions of evidence on the SDR in the two weeks after publication, working hard to 'attain the knowledge required to create a convincing critique'.[106] Nuclear deterrence had actually been ring-fenced and separate from the SDR, although decisions on the scale of the deterrent were taken.[107] The 2021 Integrated Review considers nuclear deterrence policy in more detail than any of the other reviews since the SDR, though given the change in tone, that was probably entirely appropriate.

Scrutiny of government policy decisions should be conducted by Parliament, and the Commons Defence Committee has a distinguished tradition of holding government to account on many matters. However, it takes evidence in an unclassified environment and publishes unclassified reports, so the same security inhibitions apply. Moreover, on nuclear deterrence issues, the government of the day has regularly circumvented or inhibited the Commons in their ability to apply even this level of scrutiny. The Defence Committee outrage at the government announcement of the decision to purchase Trident in July 1980, seven months before the publication of the Committee report, is illustrative: '[s]ince the House has voted, by 316 votes to 248, to endorse the choice of the Trident system, it is not for us to challenge the principle of that decision'.[108] The Committee was placed in a similar position in December 2006 when the government published *The future of the UK's strategic nuclear deterrent* and announced the Parliamentary debate would be held in March 2007.

This left the Defence Committee three months to complete a report to inform the debate: '[w]e do not express a view on the merits of retaining and renewing the UK's nuclear deterrent. Endorsing or rejecting the Government's proposals will be for the House of Commons, as a whole, to decide.'[109] Notwithstanding the limitations imposed on it by the government programme, the

Defence Committee's report concluded '[t]he Government deserves to be commended for exposing its proposal to renew the strategic nuclear deterrent to public debate and decision in Parliament, which previous Governments have not done'.[110] The report considered the timing of decisions, the scale of the UK nuclear deterrent, nuclear deterrence in the 21st century, legal and treaty obligations, and deterrent options and costs. Specifically, it did not consider whether the UK should remain a nuclear weapon state. The Defence Committee reported that it remained unclear how the government determined what constituted a minimum nuclear deterrent and advocated more clarity on what constitute UK vital interests so as not to lead to a lowering of the nuclear threshold.[111]

The debate on the White Paper in March 2007 divided in favour of the government motion 'to take the steps necessary to maintain the UK's minimum strategic nuclear deterrent beyond the life of the existing system and to take further steps towards meeting the UK's disarmament responsibilities under Article VI of the Non-Proliferation Treaty'.[112] The deputy leader of the House (Nigel Griffiths) resigned from the government in protest, followed swiftly by three other junior Labour ministers. One major topic of debate was the nature of the decision to which the motion committed the House, and whether it could expect a further substantive vote at a later stage of the procurement process. The debate was also notable for the (Labour) Foreign Secretary's response to the intervention of Labour MP and CND member Jeremy Corbyn which, she said, was 'complete and utter rubbish'.[113] In turn, Beckett was castigated by Hague (Shadow Foreign Secretary) for her own, more pliable convictions:

> It was all the more powerful coming from her, in a way, because she was a long-standing member of the Campaign for Nuclear Disarmament ... The fact that someone with her long-held views has reached the clear conclusion – in Government, and with all the information available to her – that the British nuclear deterrent must be retained, updated and replaced is in itself an indication of the powerful case for doing so.[114]

All in all, this debate was an example of British Parliamentary politics at its worst with the serious issues in the debate virtually swamped by procedural distraction, party point-scoring and personal attacks.

The press reports were mostly similarly trivial; the *Daily Express* commented on the implications of the vote for the Labour leadership: '[s]urely it is a serious indictment of Mr Blair's leadership that Labour cannot muster a parliamentary majority on a matter as crucial as defence'.[115] The *Daily Mail* and the *Daily Mirror* revelled in the government's need to rely on the Conservative opposition to carry the vote and the *Scotsman* insisted that the argument was not over.[116] The *Independent* protested that '[t]he voters have good reason to feel let down' and commented:

> The renewal of Trident ... is an issue of paramount national significance that cried out for a thorough debate. Instead, a succession of mostly lacklustre speeches preceded a vote that the Government was never going to lose ... It is regrettable that neither the Government nor David Cameron's new Conservatives could suggest anything more original than an expensive renewal of the current arrangements ... A unique chance for new thinking has been lost.[117]

Such a considered and balanced contribution towards a genuinely informed debate was notable by exception.

Since the May 2011 'Initial Gate',[118] the MOD has published an annual 'Update to Parliament' on the Dreadnought programme. These are in the form of command documents which do not require agreement or legislation. Whilst these documents are political in purpose, they are expected to conform to Parliamentary standards for rigour and accuracy and are, therefore, valuable contributions to the available authoritative information necessary to inform a public debate. The Commons Defence Committee report *Deterrence in the twenty-first century* was a very wide-ranging report on the national security strategy and the tiers of threat that document assessed.[119] However, it focused more on the use of the concept of deterrence based on conventional forces against asymmetric and terrorist threats than on the use of nuclear deterrence against state threats. Parliamentary scrutiny has been significantly inhibited by the Conservative government elected in 2019 and there is no indication that any serious rigour was applied in the limited attempts to scrutinise the 2021 Integrated Review.

Expertise in public

The expert analysis and debating of British nuclear deterrence policy in public therefore falls to non-official bodies. This is not unusual in any policy environment, but in the case of British nuclear weapons policy, there is a conspicuous absence of government participation. The 2005 Royal United Services Institute (RUSI) conference 'The future of strategic deterrence for the UK' included participation from the leading academics in the field of nuclear strategy and security studies,[120] and input from the French Foundation for Strategic Research, but no military or government participation. In 2006, the Oxford Research Group, in collaboration with the Acronym Institute for Disarmament Diplomacy, the British American Security Information Council (BASIC) and the WMD awareness Programme, formed a new initiative, 'Beyond Trident' to 'conduct new and in-depth research, foster debate in Parliament and among stakeholders, raise public awareness at all levels and create pressure for a high level non-partisan investigation and inquiry into UK nuclear weapons policy in the context of actual security needs and objectives'.[121] Its report, published in 2006, is a balanced debate which picks up on many of the key aspects of a contemporary discussion of the British nuclear deterrent, sadly lacking in the official and Parliamentary environment, with input in the form of debates by eminent academics and specialists, including serving politicians and retired officials and officers. The result was a valuable contribution to the public debate, but once more, it is bereft of official input.

The significant exception to the lack of government participation in a public debate was the 2007 RUSI debate 'Renewing Britain's independent strategic nuclear deterrent' which saw Defence Secretary Des Browne share a platform with three expert academic commentators. The debate ranged over the rationale for retention of an independent nuclear deterrent. On 'prestige', or status, Lord Browne argued that 'I would never advocate that we do this for reasons of status ... We don't rely, for our position and our status and our relationships internationally, on the fact that we are a nuclear weapons state.' Why decide now was a significant factor. The debate also considered non-proliferation, legal challenges, industrial challenges, the maintenance of CASD, a need for genuine initiatives to create the conditions for a world free of nuclear weapons and consideration of Britain's contribution to NATO. On cost, the government figures were accepted, and the focus was on the opportunity cost of those sums, but since then cost has become a highly contentious aspect of the Dreadnought project.[122]

In February 2011 BASIC established the Trident Commission, an independent, cross-party inquiry into UK nuclear weapons policy. BASIC is a small think tank consisting mostly of very senior ex-officials and politicians, supported by experienced academics, with 'one very large idea ... a world free from the threat of nuclear weapons'.[123] The Commission consisted of two previous Secretaries of State for Defence, an ex-Chief of Defence Staff, a retired UK ambassador to the UN and four eminent academic experts, including two peers. It was supported on an off-the-record basis by the Cabinet Office, FCO and MOD to ensure that government policy was explained fully, within the classification limits of the Commission's remit. The report, published in July 2014, addressed three critical questions under a national security framework:

> Should the United Kingdom continue to be a nuclear weapons state?
>
> If so, is Trident the only or best option for delivering the deterrent?
>
> What more can and should the United Kingdom do to facilitate faster progress on global nuclear disarmament?[124]

The final report was a dense read and (in an echo of the Cabinet Office quip about the *Daily Mail* version of the *Daily Telegraph* version of the 1980 Open Government document),[125] BASIC produced a two-page précis of the 'headline commission messages' to accompany it. The report, unlike official publications, attempts to deal with the critical question in Armstrong's memo to Mrs Thatcher from 1979: 'what is the nuclear deterrent for?'[126] In addressing that question, the report considered the historical legacy as it affects the present day and considered the pertinent threats. It examined three scenarios that would make retention of the strategic nuclear deterrent credible: re-emergence of a state with a significant nuclear arsenal and conventional capabilities (Russia); an existing or emerging nuclear state that enters into strategic competition with the UK; and emergence of a massive, overwhelming threat involving bio-weapons or other comparable mass destruction technologies still unknown which a state might consider explicitly using or threatening to use against the UK. The key conclusion of the Trident Commission was:

> Based upon the two key specific considerations, namely national security concerns and responsibility towards the Alliance, the Commission has come to the unanimous conclusion that the UK should retain and deploy a nuclear arsenal, with a number of caveats expressed below. Most notably, it remains crucial that the UK show keen regard for its position within the international community and for the shared responsibility to achieve progress in global nuclear disarmament.[127]

Of note is that the Commission did not see this as the final word in the debate, but merely looked to focus the debate on the weighty national security questions BASIC concluded should frame the political debate.

Public discourse tends to consist of debate conducted in the absence of official participation based on parameters and questions identified by those hostile to the government position. As detailed above, since the 2006 White Paper, the government has been very proactive in the provision of information to Parliament and the public. The 2006 White Paper itself derived a great deal from the 1980 Open Government document; the Parliamentary report into the 2011 Initial Gate was similarly informative on technical factors, and the annual reports to Parliament continue this trend. In addition, direct provision of policy information to the public such as the 2016 MOD policy paper 'UK nuclear deterrence; what you need to know' has been greatly improved, and the Parliamentary Library briefing paper 'Replacing the UK's Trident nuclear deterrent' is a very approachable description of the policy decisions and context. Although a Parliamentary, rather than government, document, the Library briefing paper describes both policy and challenging views, without drawing conclusions. It appears that current government provision of details on nuclear deterrence policy is very much better than it has been in the past, but that public engagement with those opposed to the official position remains as elusive as ever.

Public engagement

> Even when you've got Vladimir Putin waving nuclear weapons and you've got what's happening in Ukraine, the case is still not properly being made ... on the nuclear side even more so. The case goes by default because the assumption is that people will know that it is right. We've got the deterrent and Putin's rattling sabres ... we just assume that we don't have to really argue.[128]

When the 2006 White Paper was published, it formed a nucleus which anti-nuclear lobbying has focused on refuting. Much of the criticism seemed to assume that the White Paper was the output of a political essayist, rather than a carefully worked-out, extensively sourced and rigorously scrutinised piece of research by experts. There is a healthy scepticism about government pronouncements, perhaps exacerbated since the Iraq War and the Hutton inquiry, but the tendency of self-appointed critics to challenge every aspect of the derivation of nuclear policy decisions is disappointing.

Although they draw on studies with access to all of the MOD's intelligence, horizon scanning, detailed technical information and authenticated costing from industry, every government paper which concludes that Trident is the most cost-effective means of providing the deterrent has been challenged and queried. After the 2011 Initial Gate decision to proceed with the concept phase of the Dreadnought programme, the MOD conducted a further value for money study to check the predictions and, especially, the technical and financial risk in the programme. Lord Owen subsequently observed:

> if CND only read [his 2009 book *Nuclear papers*] and understand the level of debate that was going on, which was far more informed and on the nub of the issue, than anything that was going on inside CND's own discussions. I think that quite a few people have been very surprised at the level of debate that was going through.[129]

In the academic and informed debate milieu, Nick Ritchie provides probably the most comprehensive and testing intellectual critique of British deterrence policy. He has challenged the relevance of the British deterrent, the CASD policy, the decision-making on Trident's replacement and Britain's efforts towards a world free of nuclear weapons. He makes very cogent arguments reflecting personal conviction, mostly drawn from a social constructivist theoretical framework.[130] This tends to assume that the rule of law in the international environment can provide a viable sole basis for security planning. History would not support this view, having all too often had to learn Machiavelli's dictum the hard way: '[t]he fact is that a man who wants to act virtuously in every way necessarily comes to grief among so many who are not virtuous'.[131] Or, as Lord Browne put it, '[a]re we prepared to tolerate a world in which countries who care about morality lay down their nuclear weapons, leaving others to threaten the rest of the world or hold it to ransom?'[132] Governments tend to work in a realist environment, but if the public debate is already being framed in a social constructivist paradigm, it can be difficult to argue that a particular defensive capability is necessary because Britain must be capable of enforcing the rule of law, without appearing to suggest that the UK has no faith in the UN or other international bodies.

Non-engagement with CND has been deliberate on the part of government since engagement with a highly doctrinaire body such as the Campaign is regarded as pointless; it does not lead to real engagement with the debate, it merely raises the profile of the opportunity for CND to make its case. This is the 21st-century version of the 1941 debate about the 'Committee for the Abolition of Night Bombing', in which the Home Secretary responded 'I have no reason to suppose that this misguided propaganda is attracting or will attract any serious attention'.[133]

In this debate, the government is caught between engaging (and giving an otherwise low key and almost ignored debate the oxygen of publicity and longevity) and not engaging (so allowing the argument to spread within limited circles, without the countervailing policy perspective). This was the decision facing Churchill's Cabinet in the 1940s, it was Butler's experience when '[h]e exposed himself to a crossfire of questions from five accomplished controversialists who bitterly oppose the Government's basing of defence policies on the big bombs',[134] and it has been the political experience since. It is clear from the public viewing figures for *Threads* and *The Day After* that there was considerable public appetite for the subject, and the evidence above suggests that at present that appetite is being met by material that is generally partisan or simply misinterpreted. The MOD has decades of experience of not engaging with anti-nuclear campaigners; MOD personnel believe that CND argue from positions of strongly held principle and therefore cannot acknowledge the alternative perspective. (The same argument could of course be made of the MOD, but the policy papers in the public domain do make clear the degree of internal study and debate that precedes them.)

In the 1940s, Spaight was arguably a proxy voice for the official perspective in the strategic bombing debate. Such a voice was missing in the 1960s when Watkins created *The War Game*, but in the early 1980s Michael Quinlan was given extraordinary permission to engage carefully with selected opinion formers, on a non-official basis. There appears to be no government voice engaging in the 21st-century discourse, although the active participation of retired senior officials and politicians gives a welcome authority and gravitas to aspects of the debate.

> To create a climate in which defence decision-making operates sensibly, sensitively and objectively is in the interests of every citizen. It will not be achieved by avoiding the difficult moral and humanitarian issues that any defence policy inevitably raises. It can be achieved only by a far deeper public involvement in the discussion of military affairs than exists today.[135]

Little appears to have changed in the engagement of the public in defence policy generally, and nuclear deterrence policy in particular since the [then] Foreign Secretary David Owen made this point at the height of the Cold War.

The media

'Democracy cannot flourish without fair and reasoned dialogue' wrote Robin Day,[136] having interviewed every prime minister of the previous thirty-five years. He was considering the demise of the one-on-one political interview as a tool of democratic engagement and oversight:

> In the sixties and seventies major television interviews, such as those of BBC *Panorama*, were newsworthy events of much value in the political process. They attracted big headlines, verbatim news reports, fierce editorials, strong political reaction, and lively viewer response. [He would] try to ask questions which will reflect what the viewers may wish to know. But I also ask questions which the viewers ought to want asked if they knew a little more about the subject. I try to say 'what does the ordinary person want to know about this?'[137] [Political leaders may not have relished the set piece interview, but the importance was not lost on them] television has really by-passed the House of Commons in its political interviews of Ministers, not even excepting the Prime Minister and Leader of the Opposition ... Are we really willing to allow the television interview, viewed admittedly by several million people, to assume greater importance than the proceedings of the House of Commons?[138]

Robin Cook was adamant that the fault of the demise of these interviews lay with the interviewers: 'most political choices involve a trade-off between positive and negative consequences, but reducing every political interview to a one-dimensional confrontation suppresses any chance of an honest and balanced discussion of the real dilemma'.[139] Day seemed to agree with the effect: '[i]n recent years the TV interview has become more argumentative than interrogative'.[140] Andrew Marr wrote of the role of the interviewer, '[i]f your story needs to be seen more than once before it can be understood, and many do, then it will have totally failed ... You are distilling information, not packing it in. Get to the point, stick to it, know when it's finished, then end it.'[141] It is this single minded focus that agitated Cook: 'the presumption behind the badgering is that all politicians set out to evade the truth and deceive the public, which feeds cynicism with the political process'.[142]

Nott noted the difficulty of engaging on the complex ethics of nuclear deterrence in an environment where others would contest the issues in more emotive terms: '[t]o engage the emotions – as the promoters of CND know very well – is an easy task ... To argue the choices before us so as to engage the intellect is a much harder task.'[143] This remains as pertinent today as it did thirty-five years ago, or perhaps more so since there are now three generations who have no personal experience of war.

Discussing the difficulties in thinking about how to explain the basic concepts of nuclear deterrence and the ethical issues involved in a deterrence policy based on conditional willingness to launch nuclear missiles, Lord Owen noted:

> The generation that is the cross-over from the World War have grappled with this issue and it is so frightening; the numbers game in the Cold War, and it isn't just thousands, it was millions. It is almost enough to make you sick to contemplate, and therefore how do you engage on that?[144]

Notes

1. AEBC, 'Crops on trial', para. 68.
2. Peter Kellner, 'Do we want Trident? Most people want to protect defence spending – even at the expense of other services', *Prospect* (August 2015), 14.
3. *Ibid.*
4. Baylis and Stoddart, *The British nuclear experience*, 207.
5. Alec Folwell, 'Opinion formers poll: views on Trident highly politicised as Scotland vote approaches', Politics, Reputation Research, Scotland, UK. London: YouGov, 2014.
6. *Ibid.*
7. Mills, Parliamentary briefing paper 7353: 'Replacing the UK's Trident nuclear deterrent'. London: House of Commons Library/TSO, 2016.
8. *Ibid.*, p. 10.
9. Brian Cox and Jeff Forshaw, 'Education is as important to security as aircraft carriers or missile defence', *The Big Issue*, 21 November 2016.
10. UK Government, 'Trident alternatives review', 16 July 2013. London: HMSO, para. 1.
11. MOD, joint doctrine note 1/19, 'Deterrence; the defence contribution', UK MOD Doctrine and Concepts Development Centre.
12. Andrew Corbett, 'Deterring nuclear Russia in the 21st century; theory and practice', NATO Defense College Research Report. Rome: NATO Defense College, 2016.
13. I differentiate between the five nuclear weapon states which are recognised by the NPT and four nuclear-armed states (Israel, India, Pakistan and North Korea) which are not.

14 Attlee, GEN 75/3 'The atomic bomb'.
15 Fallon, 'The case for retention'.
16 Cabinet Office. *Future UK strategic nuclear deterrent force.*
17 Michael Quinlan, *Thinking about nuclear weapons; principles, problems, prospects* (Oxford: Oxford University Press, 2009), 126.
18 Wheldon to Adam, 31 December 1964.
19 Stoddart, 'Maintaining the "Moscow criterion"'.
20 HL Deb. 24 January 2007, Armed forces: nuclear deterrent. *Hansard*, Vol. 688, cols 1129–31: 1130.
21 David Owen, *Nuclear papers* (Liverpool: Liverpool University Press, 2009), 17.
22 UK Government, *Strategic Defence Review*. London: TSO, 1998, foreword para. 8 and ch. 4 paras 66–8.
23 UK Government, White Paper: *The future of the UK's strategic nuclear deterrent*: CM6994 (London: HMSO, 2006).
24 UK Government, 'Securing Britain in an age of uncertainty: the Strategic Defence and Security Review 2010' [SDSR 2010]: CM7948. (London: HMSO, 2010), foreword, p. 5 and UK Government, 'National Security Strategy and Strategic Defence and Security Review 2015: a secure and prosperous United Kingdom' [SDSR 2015]: CM 9161 (London: HMSO, 2015), p. 34 para. 4.65.
25 UK Government, *The future of the UK's strategic nuclear deterrent*, p. 13, highlight box 2.1.
26 UK Government, 'Trident alternatives review', p. 14 para. 1.9.
27 Brad Roberts, *The case for US nuclear weapons in the 21st century* (Stanford, CA: Stanford University Press, 2016), p. 35.
28 This is the logic that, in the 1960s, drove Healey and others to consider a graduated deterrence policy for NATO, which became more commonly known as flexible response.
29 I have tried to provide an informed (but not official) analysis at Andrew Corbett, '*Igitur qui desiderat pacem, praeparet bellum atomica*', *Journal of Military Ethics*, 19:4 (2020), 331–47, DOI: 10.1 080/15027570.2021.1893461.
30 UK Government, 'Global Britain in a competitive age; the Integrated Review of security, defence, development and foreign policy': CM403 (London: HMSO, 2021), 7.
31 NATO, 'The alliance's strategic concept', Brussels: NATO, 1999. Available: www.nato.int/cps/en/natohq/official_texts_27433.htm [accessed 12 March 2017].
32 UK Government, '2021 Integrated Review', 75.
33 *Ibid.*

34 *Ibid.*, 75–6.
35 UK Government, 'Trident alternatives review', para. 7.
36 Thomas C. Schelling, *Arms and influence* (New Haven, CT and London: Yale University Press, 2008), [Kindle edn], location 185.
37 UK Government, 'SDSR 2015'.
38 *Ibid.*, para. 1.3.
39 *Ibid.*, para. 1.11.
40 UK Government, 'National security capability review', London: TSO, 2018.
41 UK Government, '2021 Integrated Review', 73.
42 *Ibid.*, 75.
43 UK Government, 'Defence in a competitive age', Cm 411, 2021. London: TSO, 22.
44 Labour Party, 'The new Britain'.
45 Greenpeace, 'Trident – the UK's nuclear weapons system', [online], Greenpeace, 2006. Available: www.greenpeace.org.uk/peace/trident-the-uks-nuclear-weapons-system [accessed 15 July 2016].
46 CND, 'CND Trident mythbuster', [online], CND, 2016b. Available: www.cnduk.org/images/stories/Trident_mythbuster_2016.pdf [accessed 22 November 2016].
47 House of Commons Defence Committee 2006. 'The future of the UK's strategic nuclear deterrent: the strategic context: government response to the Committee's eighth report of session 2005–06'. Parliamentary Papers HC 1558, Session 2005–06: Parliament.
48 CND, 'CND Trident mythbuster'.
49 T. Hare, 2006. House of Commons Defence Committee, 'The future of the UK's strategic nuclear deterrent: the strategic context'. Parliamentary Papers HC 986, Session 2005–06, Ev. 35–6: Parliament.
50 HMS *Vengeance* conducted her first missile on-load in 2000, offloaded them in 2013 and reloaded in 2016 after her mid-life refit.
51 John Ainslie, 'Status of UK nuclear forces' in: Ray Acheson (ed.), *Assuring destruction forever; nuclear weapon modernization around the world*. New York: Women's International League for Peace and Freedom, 2012, pp. 67–88: 73.
52 Attlee, GEN 75/3 'The atomic bomb'.
53 Cabinet Office. GEN 75/10 'International control of atomic energy'.
54 UK Government, *The future of the UK's strategic nuclear deterrent*, Section 2 para. 2.2.
55 FCO, 'Lifting the nuclear shadow: creating the conditions for abolishing nuclear weapons', London: Foreign and Commonwealth Office, 2009, foreword p. 4.

56 UK Government, 'The road to 2010: addressing the nuclear question in the twenty-first century': Cm 7675, 2009. London: HMSO.
57 UK Government, 'Defence in a competitive age', 22.
58 S. Baldwin. HC Deb. 10 November 1932. *Hansard*, vol. 270 col. 632.
59 Air Staff, Bomber Command report 'Role and work of Bomber Command', 28 June 1942. TNA PREM 3/11/5 (166).
60 Cherwell, untitled minute, Cherwell to Churchill, 30 March 1943. TNA PREM 3/11/4(144). (Lord Cherwell was Churchill's chief scientific adviser.)
61 NATO, 'The alliance's new strategic concept', [online] Rome: NATO, 1991. Available: www.nato.int/cps/en/natohq/official_texts_23847.htm [accessed 13 September 2016].
62 NATO, 'Active engagement, modern defence', [online] Brussels: NATO, 2010. Available: www.nato.int/nato_static_fl2014/assets/pdf/pdf_publications/20120214_strategic-concept-2010-eng.pdf [accessed 23 August 2021], para. 17.
63 The author was closely involved in the drafting of this text.
64 NATO, 'Warsaw Summit communique', [online], Brussels: NATO, 2016. Available: www.nato.int/cps/en/natohq/official_texts_133169.htm?selectedLocale=en [accessed 13 September 2016], para. 54.
65 UK Government, '2021 Integrated Review', 11–12.
66 House of Commons Defence Committee 2007, Ninth report of session 2006–07, 'The future of the UK's strategic nuclear deterrent: the White Paper', London: House of Commons, 2007.
67 BASIC, 'Trident Commission concluding report; an independent, cross-party inquiry to examine UK nuclear weapons policy', London: British American Security Information Council, 2014, 6.
68 UK Government, *The future of the UK's strategic nuclear deterrent*, ch. 4 and Annex B.
69 CND, 'No to Trident' [online], London: CND, 2016c. Available: www.cnduk.org/campaigns/no-to-trident [accessed 22 November 2016].
70 £205 billion over forty-five years compares with an annual NHS budget for 2016/17 of £120.611 billion: National Health Service, 'Key statistics on the NHS', [online], NHS Confederation, 2016. Available: www.nhsconfed.org/resources/key-statistics-on-the-nhs [accessed 22 November 2016].
71 MOD, 'UK nuclear deterrence: what you need to know', London: MOD, 2016.
72 Sue Donnelly, 'CND: the story of a peace movement', *History Today*, 58 (2008).

73 CND, 'No to Trident'.
74 Healey, *The time of my life*, 242.
75 HC Deb. 28 October 1987, Defence, second day's debate. *Hansard*, vol. 121 cols 309–402: 309.
76 Cabinet Office. Minutes of a meeting of the Cabinet 8 July 1954. TNA CAB 128/27/48.
77 Healey, *The time of my life*, 302.
78 Robin Day, ... *but with respect: memorable interviews with statesmen and Parliamentarians*, London, Weidenfeld and Nicolson, 1993, 41.
79 Cabinet Office. Cabinet report by the CODP, 27 July 1954. TNA CAB 129–69–0050.
80 Armstrong to Prime Minister, A0547 'Future of the strategic deterrent', para. 6.
81 Lord Robertson, interview with Andrew Corbett, 12 November 2014.
82 *The Times*, 'Without Trident, the second division awaits: Britain's nuclear deterrent is an easy target for cuts. But the real cost has been exaggerated' (22 June 2009), p. 24.
83 L. Fox, 'Deterrence in the 21st century: speech at Chatham House, 13 July 2010'. Available: www.gov.uk/government/speeches/2010-07-13-deterrence-in-the-21st-century [accessed 23 August 2021], HMSO.
84 UK Government, 'SDSR 2010', p. 9 para. 1.4.
85 UK Government, 'SDSR 2015', p. 13 para. 2.1.
86 Cabinet Office. Report by the CODP, 27 July 1954, para. 19.
87 Cabinet Office. Unreferenced memorandum, Monckton (Cabinet Office) to Chilvers (MOD), 11 June 1956. TNA CAB 21–4054.
88 *Daily Mirror*, 'Bombs on Berlin', p. 5.
89 *Guardian*, 'Zeppelin commander on his task', p. 6.
90 S. Colllins. HC Deb. 18 June 1917. *Hansard*, vol. 94 cols 1419–21: 1419, 1420.
91 Trenchard, 'Memorandum by the Chief of the Air Staff', repr. in Webster and Frankland, *Strategic Air Offensive IV* (London: HMSO, 1961), 71–6.
92 Air Staff, letter, Street to Harris, 15 December 1943, quoted in Gray, 'The gloves will have to come off', p. 28. TNA AIR14/843.
93 Slim, *Killing civilians*, 152.
94 Macmillan and Catterall, *Macmillan diaries, Cabinet years*, 297.
95 Robert A. Fitch, 'An experimental Christian ethics', *The Journal of Religion*, 20 (1940), 325–39: 329.
96 *Ibid.*, 336.
97 Herbert Butterfield, *History and human relations*, London: Collins, 1951.

98 Collins. HC Deb. 18 June 1917.
 99 This idea bears some similarity to Baylis's 'permanent dialogue'; John Baylis, *Nuclear ethics, realism and utopianism: the 'permanent dialogue' revisited*, Swansea: University of Wales, 2002.
100 Nott to PM, 'Trident, public attitudes'.
101 Author's personal experience; ACSC 1999–2000, SSBN command 2003–7, MOD 2009–11.
102 CND, 'CND free teaching resources', [online], CND, 2016a. Available: www.cnduk.org/information/peace-education/teaching-resources [accessed 22 November 2016].
103 Warnock, 'Moral thinking and government policy', 513.
104 Peter M. Sandman, 'Explaining risk to non-experts: a communications challenge', *Emergency Preparedness Digest* (1987a), 25–9: 25.
105 Robertson, interview with Corbett 2014.
106 Bruce George, 'Political perspectives on the outcome of SDR: the House of Commons Select Committee report', *The RUSI Journal*, 143 (1998), 26.
107 UK Government, 1998 *Strategic Defence Review*, supporting paper 5, paras 6–9.
108 House of Commons Defence Committee 1981, 'Fourth report from the Defence Committee session 1980–81', para. 1.
109 House of Commons Defence Committee, Ninth report: 'The White Paper', para. 3.
110 *Ibid.*, para. 186.
111 *Ibid.*, paras 64 and 81.
112 HC Deb. 14 March 2007, Trident. House of Commons debates, Parliamentary Business home page: Houses of Parliament. Available: https://publications.parliament.uk/pa/cm200607/cmhansrd/cm070314/debtext/70314-0005.htm [accessed 23 August 2021].
113 *Ibid.*, col. 305.
114 *Ibid.*, col. 310.
115 *Daily Express*, 'Can our defence ever be safe in Labour hands?' (15 March 2007), p. 10.
116 *Daily Mail*, 'The great rebellion; 95 Labour MPs join revolt but Tory votes rescue Blair' (15 March 2007), p. 4; *Daily Mirror*, '95 rebels; Blair humiliated as he's forced to rely on the Tories to win nuclear missiles battle' (15 March 2007), p. 6 and *Scotsman*, 'Trident argument is not over' (15 March 2007), p. 29.
117 *Independent*, 'The voters have good reason to feel let down' (15 March 2007), p. 40.

118 Initial Gate is the first Parliamentary approval for significant investment in large defence procurement projects. It authorises the assessment of different options to meet defined key user requirements within the authorised cost and time parameters. For more detail, see MOD Joint Service Publication 507, 'Investment appraisal and evaluation'; available: https://assets.publishing.service.gov.uk/government/uploads/system/uploads/attachment_data/file/275548/JSP507_Part_1_U.pdf [accessed 23 August 2021], ch. 2.
119 House of Commons Defence Committee 2013, *Eleventh report of session 2013: Deterrence in the twenty-first century*, vol. 1, London: House of Commons, 2013.
120 RUSI, 'The future of the UK strategic deterrent', London: RUSI, 2005.
121 Ken Booth and Frank Barnaby, *The future of Britain's nuclear weapons: experts reframe the debate*, Oxford: Oxford Research Group, 2006.
122 Lee Willett, 'Renewing Britain's independent strategic nuclear deterrent: a debate', [online], RUSI, 2007. Available: https://rusi.org/commentary/renewing-britain's-independent-strategic-nuclear-deterrent-debate [accessed 20 November 2016].
123 BASIC, 'Trident Commission concluding report', iv.
124 *Ibid.*
125 Norbury to Whitmore, MO18/1/1 'Polaris successor'.
126 Armstrong to PM, A0547 'Future of the strategic deterrent'.
127 BASIC, 'Trident Commission concluding report': 6.
128 Robertson, interview with Corbett 2014.
129 Lord Owen, interview with Andrew Corbett (16 April 2015).
130 See: Nick Ritchie, *Trident: the deal isn't done; serious questions remain unanswered*, Bradford: Bradford University, 2007; Nick Ritchie, 'Deterrence dogma? Challenging the relevance of British nuclear weapons', *International Affairs*, 85 (2009), 81–98; Nick Ritchie and Paul Ingram, 'A progressive nuclear policy: rethinking continuous at-sea deterrence', *RUSI Journal*, 155 (2010); and Nick Ritchie, *A nuclear weapons-free world?: Britain, Trident and the challenges ahead*, London: Palgrave Macmillan, 2012.
131 Niccolò Machiavelli, *The Prince*, London: Penguin, 1513; 1995: 48.
132 Des Browne, *The United Kingdom's nuclear deterrent in the 21st century*, London: TSO, 2007.
133 HC Deb. 27 November 1941. Hansard, vol. 376 cols 886–7.
134 *The Times*, 'Nuclear tests challenge' (1 April 1958).
135 David Owen, *The politics of defence*, London: Jonathan Cape, 1972, 235.

136 Day, ... *but with respect*, 3.
137 *Ibid.*, 2, 293.
138 Brigadier Sir John Smythe VC, MP, quoted in *ibid.*, 34.
139 Robin Cook, *The point of departure*, London: Simon and Schuster, 2003, 57.
140 Day, ... *but with respect*, 6.
141 Andrew Marr, *My trade*, London: Pan Macmillan, 2004, 267.
142 Cook, *The point of departure*, 57.
143 HC Deb. 3 March 1981, Nuclear deterrence. *Hansard*, vol. 1000, cols 137–224: 137.
144 Owen, interview with Corbett, 2015.

Conclusion
Dirty hands and the supreme emergency

I have argued that there has been a sustained reluctance amongst British governments to engage in public dialogue on strategic nuclear deterrence policy. Some of the factors in the decision-making process for that policy have their roots in the arguments about reprisals for air raid attacks during the First World War. That experience and similar factors were significant for decision-making on strategy and public presentation of strategy during the bombing campaign of the Second Word War. Strategic factors such as cost, capability and industrial capacity have had a significant impact on the choice of systems employed by the British strategic nuclear forces, including ultimately driving the decisions to procure Polaris and Trident, but they have had limited impact on the evolution of the uniquely British understanding of nuclear deterrence.

Every nuclear weapon state has come by their nuclear weapons through a different dynamic and has a different understanding of nuclear deterrence theory and nuclear strategy to any other state. British theories and motives cannot readily be determined using the same logic as other nuclear weapon states may pursue. Due to decades of reticence in this policy area, within and outside government, there is a need for education in the lexicon of that theory and strategy at all levels in Whitehall and Westminster. This would inform a profound shift in the leadership exercised in the nuclear policy area if the hard decisions required to ensure the maintenance of the strategic nuclear deterrent are not to be derailed by ineptitude or simple ignorance.

Moral factors have had a profound effect on the nature of Britain's nuclear deterrent. Britain has been a reluctant nuclear power; always trying to balance its realist understanding of the utility of nuclear weapons with an idealist aspiration to find a way to create the conditions for a world free of them, although that concern appears less prevalent in the 2021 Integrated Review. The two are usually in tension, most evidently during periods where decisions on the future of nuclear weapons systems are required, but are seldom discussed.

Technical factors have affected the decisions taken about the nature and scale of the British nuclear deterrent. The perceived scale of destructive capability necessary to deter the Soviet Union drove the decision to procure systems as capable as Polaris and subsequently Trident, and limited national industrial capacity and cost drove the decision to seek those systems from the USA. A similar balance of technical and moral factors pertains today; since the 2006 White Paper, the British government has been more open about the technical factors driving technical and system decisions, but as guarded as ever about the moral factors which influence the decision to remain a nuclear weapon state.

The British government has used different means of dealing with policy development in challenging ethical areas, some of which are considered above. In the case of nuclear weapons policy, governments have invariably reverted to reticence and avoidance of the issue in public debate, and that has been the crux of this analysis: why?

In a contemporary, rights-based ethical paradigm, the concept of war as a tool for political objectives is nearly untenable. In the realist paradigm in use by statesmen, war is justified within certain parameters. The evolution of British ethical thinking after the German air attacks of the First World War and the Allied strategic bombing campaign of the Second World War pushed the boundaries of what could be considered acceptable in a war for national survival, and led subsequently to the evolution of the concept of 'supreme emergency'. This concept excuses, although it does not justify, the killing of non-combatants under certain circumstances. It leaves those involved with 'dirty hands'. Even so, the supreme emergency concept is incompatible with modern, or revisionist, ethical models.

The nature of contemporary media is such that complex issues tend to be reduced to simple, catchy phrases; soundbites. These cannot convey the range of factors required to make decisions about fundamental moral questions such as human embryology, genetic modification of crops or nuclear deterrence policy. In these ethically challenging areas, the public are uncomfortable deferring to experts, and tend to want to know more for themselves. But modern media are not equipped to provide the necessary environment for detailed and considered debate. Nor is this common any longer in any Parliamentary environment, increasingly including select committee evidence sessions, although it can be found in specialist environments such as think tanks. As a result of feeling unable to articulate the complex ethical factors associated with these policy areas in a suitable public environment, governments have tended to avoid occasions where *extempore* intervention might be necessary, and official intervention has tended to be either by proxy, formulaic or more usually not present at all.

The current process for procuring Dreadnought is symptomatic of this lack of engagement. Since the 2006 White Paper, however, the Labour government, the 2010–15 Coalition and, it appears, the 2015 Conservative government have addressed this to a limited extent with the formulation of the Cabinet Office Communications Steering Group, and there is a coherent ongoing process of conveying official nuclear deterrence policy through government papers and publications. This is a significant step forward, although it still concentrates on the technical factors. The announcement of the uplift in warhead stockpile numbers and the return to strategic ambiguity about deployed and operational warheads and missiles in the 2021 Integrated Review is atypical of this focus on the technical to the detriment of public understanding of the decision.

The cornerstone of British defence and security policy remains nuclear deterrence. The present public debate about the Dreadnought class is essentially sterile and tends to be in a context set by anti-nuclear lobbyists, with the reactive official view appearing defensive. Technical discussions about cost and capability have become the primary area of argument because the underlying rationale for why nuclear deterrence is important to the UK is not understood well, if at all; and is certainly not well articulated. There is more risk now than ever before that future decisions will be made purely on financial grounds and the nuclear deterrent programme cancelled almost by accident, or without understanding the strategic ramifications such a decision entails.

To enable nuclear-policy decision-making to operate in the more open style of 21st-century British government is in the interests of every one of us. This will not be achieved by perpetuating the historic avoidance of difficult ethical issues that it raises. It will be achieved only through better education of 'experts' and political leaders in deterrence ethics, better communication of those expert views to the public, and better education of the population at large in order to be able to assess competing ethical paradigms and claims (including government assertions). The experience of handling the coronavirus pandemic suggests that this coherent, detailed engagement remains a challenging proposition in the 2020s.

Bibliography

Primary sources

Full reference and access details for primary sources are attached at the relevant endnote in each chapter.

Films/broadcasts

Doom town, 1955. Newsreel. Directed by Pathé. London: Pathé.
Nuclear war in Britain; home front civil defence films 1951–1987, 2010. Strikeforce TV.
'Reith Lecture: Russia, the atom and the West', 1957. Radio. Given by Kennan, G. F. London: BBC.
Things to Come, 1936. Directed by Menzies, W. C. London: London Film Productions.
Threads, 1984. Directed by Jackson, M. London: BBC.
War Game, The, 1965. Directed by Watkins, Peter. London: BBC.
When the Wind Blows, 1986. Directed by Murakami, J. T. London: Film Four International and British Screen Productions.

Interviews

Lord David Owen, interview with Corbett, A., 16 April 2015.
Lord George Robertson, interview with Corbett, A., 12 November 2014.
Watkins, Peter, interview with Corbett, A., 27 September 2012.

Treaties and conventions

Hague Conference, 'Convention (II) with respect to the laws and customs of war on land and its annex: regulations concerning the laws and customs of war on land' (1907). Available: www.icrc.org/applic/ihl/ihl.

nsf/Treaty.xsp?action=openDocument&documentId=CD0F6C83F96FB459C12563CD002D66A1 [accessed 18 December 2013].

Hague Conference, 'Rules concerning the control of wireless telegraphy in time of war and air warfare' (1923). Available: www.icrc.org/applic/ihl/ihl.nsf/Article.xsp?action=openDocument&documentId=3876F3A2997A8103C12563CD00518519 [accessed 18 December 2013].

Charter of the United Nations (1945). San Francisco: UN. Available: www.un.org/en/about-us/un-charter/full-text [accessed 23 August 2021].

Parliamentary debates and related papers

Hansard entries follow the *Hansard* webpage forms:

HC: House of Commons;
HL: House of Lords.

HC Deb., 19 April 1917, *Hansard* vol. 92 col. 1815.
HC Deb., 24 April 1917, *Hansard* vol. 92 cols 2224–5.
HC Deb., 7 June 1917, *Hansard* vol. 94 col. 353W.
HC Deb., 14 June 1917, *Hansard* vol. 94 col. 1136.
HC Deb., 18 June 1917, *Hansard* vol. 94 cols 1419–21.
HC Deb., 25 June 1917, *Hansard* vol. 95 cols 14–5 14.
HC Deb., 11 July 1928, *Hansard* vol. 71 cols 963–86.
HC Deb., 10 November 1932, *Hansard* vol. 270 col. 632.
HC Deb., 28 April 1937, *Hansard* vol. 323 cols 312–9.
HC Deb., 3 June 1938, *Hansard* vol. 336 cols 2411–3.
HC Deb., 21 June 1938, *Hansard* vol. 337 cols 937–8.
HC Deb., 3 October 1938, *Hansard* vol. 339 cols 40–162.
HC Deb., 24 July 1941, *Hansard* vol. 373 cols 1051–2.
HC Deb., 27 November 1941, *Hansard* vol. 376 cols 886–7.
HC Deb., 11 March 1943, *Hansard* vol. 387 cols 922–62.
HC Deb., 31 March 1943, *Hansard* vol. 388 col. 155.
HC Deb., 21 August 1945, Atomic Energy Committee. *Hansard* vol. 413 cols 442–3.
HC Deb., 1 March 1948, Defence. *Hansard* vol. 448 cols 87–160.
HC Deb., 12 May 1948, *Hansard* vol. 450 col. 2117.
HC Deb., 25 October 1962, *Hansard* vol. 664 cols 1053–64.
HC Deb., 21 March 1978, *Hansard* vol. 946 cols 1313–15.
HC Deb., 24 January 1980, *Hansard* vol. 977 cols 672–784.
HC Deb., 15 July 1980, Statement on the strategic nuclear deterrent. *Hansard* vol. 988 cols 1235–51.
HC Deb., 3 March 1981, Nuclear deterrence. *Hansard* vol. 1000 cols 137–224.

HC Deb., 11 March 1982, Trident missile programme. *Hansard* vol. 19 cols 975–86.
HC Deb., 28 October 1987, Defence, second day's debate. *Hansard* vol. 121 cols 309–402 309.
HC Deb. 14 March 2007, Trident. House of Commons Debs, Parliamentary Business home page: Houses of Parliament. Available: https://publications.parliament.uk/pa/cm200607/cmhansrd/cm070314/debtext/70314-0005.htm [accessed 23 August 2021].
HL Deb., 29 April 1937, *Hansard* vol. 105 cols 84–92.
HL Deb., 3 October 1938, *Hansard* vol. 110 cols 1297–366.
HL Deb., 13 September 1939, *Hansard* vol. 114 cols 1047, 1050–52.
HL Deb., 17 March 1948, *Hansard* vol. 154 cols 863–926.
HL Deb., 24 March 1983, Defence Secretariat 17 and 19. *Hansard* vol. 440 cols 1226–30.
HL Deb., 24 January 2007, Armed forces: nuclear deterrent. *Hansard* vol. 688 cols 1129–31.
House of Commons Defence Committee, 'The future of the UK's strategic nuclear deterrent: the strategic context: government response to the Committee's Eighth report of session 2005–06'. Parliamentary Papers HC 1558, Session 2005–06: Parliament.
House of Commons Defence Committee, Ninth report of session 2006–07: 'The future of the UK's strategic nuclear deterrent: the White Paper', 2007. London: House of Commons.
House of Commons Defence Committee, Eleventh report of session 2013: 'Deterrence in the twenty-first century', vol. 1, 2013. London: House of Commons.
Hare, T. 2006. Defence Committee, 'The future of the UK's strategic nuclear deterrent: the strategic context'. Parliamentary Papers; HC 986, Session 2005–06, Ev. 35–6: Parliament.
Mills, C. 2016. Briefing Paper 7353: 'Replacing the UK's "Trident" nuclear deterrent'. In: House of Commons Library. London: TSO. Available: https://commonslibrary.parliament.uk/research-briefings/cbp-7353/ [accessed 12 September 2019].

Cabinet Office papers

Minutes of a meeting of the Imperial War Cabinet, 3 April 1917. TNA CAB 23/40/07.
Minutes of a meeting of the Imperial War Cabinet, 12 April 1917. TNA CAB 23/40/09.
Minutes of a meeting of the War Cabinet, 14 June 1917. TNA CAB 23-4-10.

Bibliography 231

Letter from Mr A Baker (Hertford) to Sec., War Cabinet with resolution of a m[ee]t[in]g held in Hertford on 26 June [1917]. TNA CAB 24/17/75.

Letter from Mr GJ Allen (Croydon) to Sec., War Cabinet with resolution of a m[ee]t[in]g held in Croydon (GT-1373 dated 26 June 1917). TNA CAB 24/17/74.

Minutes of a meeting of the War Cabinet, 7 July 1917. TNA CAB 23/3/26.

Letter from Dr Macnamara MP; petition on reprisals from constituents of Camberwell 9 July 1917. TNA CAB 24/17/76.

Second report of the Prime Minister's Committee on Air Organisation and Home Defence against Air Raids. 17 August 1917 (the Smuts Report). TNA AIR 1/ 515/16/3/83.

Minutes of a meeting of the War Cabinet, am 2 October 1917. TNA CAB 23/4/17.

Minutes of a meeting of the War Cabinet, pm 2 October 1917. TNA CAB 23/4/18.

Conclusions of a meeting of the Cabinet, 28 April 1937. TNA CAB 23/88/07.

Conclusions of a meeting of the Cabinet on 5 May 1937. TNA CAB 23/88/08.

Minutes of a meeting of the War Cabinet, Friday 10 May 1940. TNA CAB 69/1.

Defence Committee (Operations), Minutes of a meeting of the War Cabinet, Tuesday 24 September 1940. TNA CAB 69/1.

Defence Committee (Operations), Minutes of a meeting of the War Cabinet, 13 January 1941. TNA CAB 69/2.

Defence Committee (Operations), Minutes of a meeting of the War Cabinet, 3 July 1941. TNA CAB 69/2.

War Cabinet; Chiefs of Staff Committee. Minutes of a meeting held on 14 July 1941. TNA CAB 79/12.

Defence Committee (Operations), Memorandum on bombing policy. DO(42)14, 9 February 1942. TNA CAB 69-4.

Defence Committee (Operations), Memorandum: Estimation of bombing effect. DO(42)38, 9 April 1942. TNA CAB 69/2.

Defence Committee (Operations), Memorandum: Estimation of bombing effect. DO(42)39, 13 April 1942. TNA CAB 69/4.

Report by Mr Justice Singleton, 'The bombing of Germany', May 1942. TNA PREM 3/11/24 (112).

Attlee, C. 'The atomic bomb: memorandum by the Prime Minister'. GEN 75/1, 28 August 1945. TNA CAB 130/3.

Attlee, C. GEN 75/3, 'The atomic bomb: letter from the Prime Minister to President Truman', 25 September 1945. TNA CAB 130/3.

GEN 75/10 'International control of atomic energy; report by officials', 29 October 1945. TNA CAB 130/3.

GEN 75/7th Meeting: Note of a meeting of ministers held at No. 10 Downing Street on Thursday 1 November 1945. TNA CAB 130/2.

DO(46) 5th Meeting, Cabinet Defence Committee, Minutes of a meeting held at No. 10 Downing Street on Friday 15 February 1946. TNA CAB 131/1.

DO(46) 17th Meeting, Cabinet Defence Committee, Minutes of a meeting held at No. 10 Downing Street on Monday 27 May 1946. TNA CAB 131/1.

Atomic Energy Commission of the United Nations; second meeting; verbatim record. 19 June 1946. TNA CAB 130/3.

GEN 75/37 International control of atomic energy: the United States and Soviet proposals, 4 July 1946. TNA CAB 130/3.

GEN 163/1st Meeting, Note of a meeting of ministers held at No. 10 Downing Street on Wednesday 8 January 1947. TNA CAB 130/16.

Untitled memorandum, Churchill (PM) to Bridges (Treasury), 8 December 1951. TNA PREM 11/297.

Note of a meeting in the Cabinet Office, GEN 465, 12 March 1954. TNA CAB 130/101.

Minutes of a meeting of the Cabinet, 22 March 1954. TNA CAB 128/27/21.

Note of a meeting of ministers, GEN 464, 13 April 1954. TNA CAB 130/101.

Minutes of a meeting held at 10 Downing Street on Thursday 24 April 1952. TNA ADM 116/6087.

Minutes of a meeting of the Committee on Defence Policy, DP(54) 2nd Meeting, 19 May 1954. TNA CAB 134/808.

Memorandum by the Chiefs of Staff for the Defence Policy Committee and Cabinet: 'United Kingdom defence policy'. 31 May 1954. TNA CAB 129/69.

Minutes of a meeting of the Cabinet, 8 July 1954. TNA CAB 128/27/48.

Cabinet Report by the Committee on Defence Policy. 27 July 1954. TNA CAB 129-69-0050.

D(55)17 'The defence implications of fall out from a hydrogen bomb' (the Strath Report). 1955. TNA CAB 21-4054.

Unreferenced memorandum Monckton (Cabinet Office) to Chilvers (MOD), 11 June 1956. TNA CAB 21-4054.

BND(SG)(60)3 12 April 1960. TNA DEFE 10/665.

Conclusions of a meeting of the Cabinet, Thursday 28 July 1960. TNA CAB/128/34.

Conclusions of a meeting of the Cabinet held at Admiralty House on Thursday 15 September 1960. TNA CAB/128/34.

Bibliography 233

CC(62) 61st Conclusions, Conclusions of a meeting of the Cabinet, 23 October 1962. TNA CAB/128/36.

CC(62) 62nd Conclusions, Conclusions of a meeting of the Cabinet, 25 October 1962. TNA CAB/128/36.

MISC 17/4 Cabinet Defence Policy, Minutes of a meeting at Chequers, 22 November 1964. TNA CAB 130/213.

Memorandum, Sir Burke Trend to Wilson, 8 October 1965, The National Archives (TNA) PREM 13/139.

MISC 17/4 Defence Review: report to ministers by an official committee (chairman Sir B Trend) of the Cabinet Defence and Overseas Policy Committee, 8 November 1965. TNA CAB 130/213.

MISC 17/8 Defence policy: Record of a meeting at 10 Downing Street of ministers, service chiefs and senior officials, 13 November 1965. TNA CAB 130/213.

Defence Review: Minute by Mr Walker to Prime Minister on the issue of 'whether we are an island off the north west corner of Europe, or a World Power', 23 November 1965. TNA PREM 13/216.

CC(68)3 Public expenditure: post devaluation measures: Cabinet conclusions on withdrawal from east of Suez. 4 January 1968. TNA CAB 128/43.

Foreign Office Telegram No. 554: Text of Mr Wilson's reply to President Johnson's letter of 11 January 1968. 15 January 1968. TNA PREM 13/1999.

A05828 Military nuclear issues. 25 October 1977. TNA PREM 16/1564.

Conclusions of a ministerial meeting held at No. 10 Downing Street on Friday 28 October 1977 at 0945. TNA PREM16/1564.

A06085 Nuclear matters. 28 November 1977. Loose minute from Cabinet Secretary to PM, TNA PREM16/1564.

Factors relating to the further consideration of the future of the United Kingdom deterrent, 7 December 1978. In: Cabinet Office (ed.). TNA DEFE 19/275.

Memorandum (Hunt (Cabinet Secretary) to Prime Minister), A09454 The future of the deterrent. 4 May 1979. TNA 19/14.

Memorandum (Hunt (Cabinet Secretary) to Prime Minister), A09588 The future of the deterrent. 18 May 1979. TNA PREM 19/14.

Memorandum Armstrong (Cabinet Secretary) to Prime Minister, A0500 Future of the strategic deterrent. 29 October 1979. TNA PREM 19/14.

Memorandum Armstrong (Cabinet Secretary) to Prime Minister, A0547 Future of the strategic deterrent. 2 November 1979. TNA PREM 19/14.

Memorandum Armstrong (Cabinet Secretary) to Prime Minister, A0548. 2 November 1979. TNA PREM 19/14.

Memorandum Armstrong (Cabinet Secretary) to Prime Minister, A0940 Modernisation of NATO's long range theatre nuclear forces and the replacement of Polaris. 12 December 1979. TNA PREM 19-0159.

A01003 Note of a meeting in the Oval Office, the White House, Washington DC, on Monday 17th December 1979. TNA PREM 19-0159.

Memorandum Wade-Gery (Cabinet Office) to Armstrong (Cabinet Secretary), B05909 Polaris replacement. 11 February 1980. TNA PREM 19-0159.

Memorandum Cabinet Secretary to Cabinet, OD(80)23 Defence and Overseas Policy Committee civil home defence. 18 March 1980. TNA CAB 148/190.

OD(80) 9th Meeting of Cabinet Defence Overseas Policy Committee, Minutes, Thursday 20 March 1980. TNA CAB 148/189.

Unreferenced memorandum, Polaris successor: memorandum. 17 June 1980. TNA PREM 19/417.

OD(80) 18th meeting, Cabinet Defence and Overseas Policy Committee, Meeting minutes, Tuesday 8 July 1980. TNA CAB 148/189.

Memorandum Armstrong (Cabinet Secretary) to Prime Minister, A09119 The United Kingdom strategic deterrent; missile processing. MISC 7(82)4 27 July 1982. TNA PREM 19/0695.

Letter Goodall (Cabinet Office) to FCO, Meeting with the Roman Catholic Bishops of England and Wales: nuclear defence issues, 14 January 1983. Margaret Thatcher Foundation Archive: Thatcher MSS (Churchill Archive Centre): THCR 1/4/7 f4. Available: https://www.margaretthatcher.org [accessed 12 September 2019].

CC(83) 14th Conclusions, Conclusions of a meeting of the Cabinet held at 10 Downing Street, Thursday 28 April 1983. TNA CAB 128/76/14.

Government publicationsAGRICULTURE AND ENVIRONMENT BIOTECHNOLOGY COMMISSION 'Crops on trial; a report by the AEBC', September 2001. London: HMSO.

BR 4005, *Manual of Naval Security*, vol. 1. TNA ADM 234/1134.

COMMITTEE OF IMPERIAL DEFENCE, 'Appreciation of the situation in the event of war against Germany' (COS 764), 13 September 1938. TNA CAB 53/41.

Committee of Imperial Defence, 'War plans: report by the joint planning sub-committee' (COS 781), 25 October 1938. TNA CAB 53/41.

DEFENCE COMMITTEE, Defence Committee alternative draft report, March 1981. London: HMSO.

Defence Committee, 'Fourth report from the Defence Committee session 1980–81.' London: HMSO.

DEPARTMENT FOR ENVIRONMENT, FOOD AND RURAL AFFAIRS, 'The GM dialogue: government response', 9 March 2004. London: HMSO.

Department for Environment, Food and Rural Affairs, 'The GM public debate: lesson learned from the process', March 2004. London: HMSO.

DEPARTMENT OF HEALTH AND SOCIAL SECURITY, Press release 82/230 – 'Government enquiry into human fertilisation', 23 July 1982. TNA FD7-2307.

FCO, 'Lifting the nuclear shadow: creating the conditions for abolishing nuclear weapons', 2009. London: Foreign and Commonwealth Office. Available: http://carnegieendowment.org/files/nuclear-paper.pdf [accessed 12 September 2019].

FOOD STANDARDS AGENCY, 'Consumer views of GM food', 2003. London: HMSO.

HOME OFFICE, 'Protect and survive: civil defence manual of basic training', 1950. London: HMSO.

Home Office, Civil Defence Handbook No. 10: 'Advising the householder on protection against nuclear attack', 1963. London: HMSO.

NHS, Key statistics on the NHS [online] 2016. NHS Confederation. Available: www.nhsconfed.org/resources/key-statistics-on-the-nhs [accessed 22 November 2016].

UK GOVERNMENT, Defence and Security Media Advisory (DSMA) Notice 02: 'Nuclear and non-nuclear weapons and equipment'. Available: www.dsma.uk/notice/nuclear-non-nuclear-weapon-systems-equipment/ [accessed 5 August 2021].

UK Government, 'National Security Strategy and Strategic Defence and Security Review 2015: a secure and prosperous United Kingdom'. CM 9161. London: HMSO. Available: https://assets.publishing.servi ce.gov.uk/government/uploads/system/uploads/attachment_data/ file/478933/52309_Cm_9161_NSS_SD_Review_web_only.pdf [accessed 20 July 2016].

UK Government, 'The road to 2010: addressing the nuclear question in the twenty-first century', Cm 7675, 2009. London: HMSO. Available: https:// assets.publishing.service.gov.uk/government/uploads/system/uploads/ attachment_data/file/238560/7675.pdf [accessed 24 August 2021].

UK Government, 'Securing Britain in an age of uncertainty: the Strategic Defence and Security Review 2010', Cm 7948. London: HMSO. Available: https://assets.publishing.service.gov.uk/government/uploads/ system/uploads/attachment_data/file/62482/strategic-defence-security-review.pdf [accessed 12 September 2019].

UK Government, 'Trident alternatives review', 16 July 2013. London: HMSO.

UK Government, 'The future United Kingdom strategic nuclear deterrent force', 1980. London: HMSO.

UK Government, 'Strategic Defence Review', 1998. London: TSO. Available: http://www.mod.uk/NR/rdonlyres/65F3D7AC-4340-4119-

93A2-20825848E50E/0/sdr1998_complete.pdf [accessed 12 September 2019].

UK Government, 'The future of the UK's strategic nuclear deterrent: the White Paper', Cm 6994, 2006. London: HMSO. Available: https://assets.publishing.service.gov.uk/government/uploads/system/uploads/attachment_data/file/27378/DefenceWhitePaper2006_Cm6994.pdf [accessed 12 September 2019].

UK Government, 'Defence in a competitive age', Cm 411, 2021. London: TSO.

Warnock, M. 'Report of the Committee of Inquiry into human fertilization and embryology', 1984. London: HMSO.

Defence papers

AIR DIVISION BAFO 1946. 'Report on German flak towers', Flak Disarmament Branch. TNA AIR 55/ 158.

AIR STAFF. Original paper on objectives, 3 September 1917. TNA AIR 1/462/15/312/121.

Air Staff. Operational Requirement OR229. TNA AIR20/2240.

Air Staff. Bombing directive, 9 July 1941. Reproduced in Webster and Frankland (1961), *The Strategic Air Offensive against Germany 1939–1945*, vol. IV, London, HMSO, pp. 135–9.

Air Staff. Bomber Command report, 'Role and work of Bomber Command', 28 June 1942. TNA PREM 3/11/5 (166).

Air Staff. Bombing directive, 14 February 1942. Reproduced in Webster and Frankland (1961), *The Strategic Air Offensive against Germany 1939–1945*, vol. IV, pp. 143–5.

Air Staff. Letter, Harris to Portal, ATH/DO/4 (A), 1 November 1944. DEAN 02/10 (ATH/DO/4) Liddell Hart Archives.

Air Staff. Letter, Harris to Portal, ATH/DO/4 (J), 18 January 1945. DEAN 02/10 (ATH/DO/4) Liddell Hart Archives.

Air Staff. Letter, Street to Harris, CS.21079/43, 15 December 1943. TNA 14/843.

CHERWELL, Untitled minute, Cherwell to Churchill, 30 March 1943. TNA PREM 3/11/4(144).

FCO, Memorandum, Carrington (FCO) to Prime Minister PM/80/ 45, Polaris successor, 12 June 1980. In: FCO (ed.). TNA PREM 19/417.

JOINT INTELLIGENCE COMMITTEE, 'Effect of bombing policy', report by the Joint Intelligence Sub-Committee JIC(42)117(D), 6 April 1942. TNA CAB 79/20/16.

MADDEN, C. Note by the Chief of Naval Staff for the Chiefs of Staff Sub-Committee on the memorandum of the Chief of the Air Staff,

21 May 1928. Reproduced in Webster and Frankland (1961), *The Strategic Air Offensive against Germany 1939–1945*, vol. IV, London, HMSO, pp. 81–3.

MILNE, G. Note by the Chief of the General Staff for the Chiefs of Staff Sub-Committee on the memorandum of the Chief of the Air Staff, 21 May 1928. Reproduced in Webster and Frankland (1961), *The Strategic Air Offensive Against Germany 1939–1945*, vol. IV, London, HMSO, pp. 76–81.

MOD, Admiralty 1945a. ACNS(W), 15 August 1945. TNA ADM 1/117259.

MOD, Admiralty 1945b. Director Plans, 2 September 1945. TNA ADM 1/117259.

MOD, Memorandum, Macmillan (MOD) to Churchill (Prime Minister), 8 December 1954. TNA DEFE 13/45.

MOD, Unreferenced memorandum, Lord Selkirk to Lord Hailsham, 1 January 1958. TNA/ ADM/ 1/ 27375.

MOD, BND(SG)(59)1, British Nuclear Deterrent Study Group – minutes of a meeting held Monday 16 July 1959. In: MOD (ed.). TNA DEFE 10/665.

MOD (air staff), VCAS British Nuclear Study Group Monday (BND(SG)(59)8, 19 October 1959. In: MOD, A. S. (ed.). TNA AIR 2/13707.

MOD, BND(SG)(59)15, British Nuclear Deterrent Study Group – draft interim report, 30 October 1959. TNA DEFE 10/665.

MOD (air staff), Unreferenced note, Sec. BNSG to Sir Solly Zuckerman, 'Future deterrent policy', 26 September 1960. TNA DEFE 19/11.

MOD, 'The United Kingdom defence review', draft *aide mémoire* by HMG for discussion in Washington and Canberra, January 1966. TNA DEFE 13/477.

MOD, Memorandum from PPS to Prime Minister to Cabinet Secretary, 'UK nuclear deterrent', 30 January 1978. TNA PREM 16/1654.

MOD, Parliamentary Question handling brief, 21 March 1978. TNA PREM 16/1564.

MOD, Memorandum, Pym to Thatcher, MO 13/1/34, 'NATO long-range theatre nuclear forces', 5 July 1979. TNA DEFE 25/335.

MOD, Memorandum DUS(P) 436/79, 'Long-range theatre nuclear forces', 6 July 1979. TNA DEFE 24/2122.

MOD, Memorandum, Blelloch to DUS(P) MOD, 6 August 1979. TNA DEFE 24/2122.

MOD, Loose minute DUS(P) 619/79, 'TNF modernisation – visit by Mr David Aaron', 22 October 1979. TNA PREM 19/14.

MOD, Memorandum, Pym (MOD) to Prime Minister, MO 18/1/1, 'Polaris successor', 10 June 1980. In: MOD (ed.). TNA PREM 19/417.

MOD, Unreferenced personal memorandum, Prime Minister to Pym (MOD), 'Polaris successor', 11 June 1980. TNA PREM 19/417.

MOD, Unreferenced memorandum, Norbury (PUS at MOD) to Whitmore (Cabinet Office), MO18/1/1, 'Polaris successor', 23 June 1980. TNA PREM 19/417.

MOD, Unreferenced memorandum, Whitmore (Cabinet Office) to Norbury (MOD), 'Polaris successor', 7 July 1980. TNA PREM 19/417.

MOD, Personal memorandum, Nott (MOD) to Prime Minister, MO18/1/1, 'Trident, public attitudes', 2 February 1981. TNA PREM 19/555.

MOD, Letter, Nott (MOD) to Thatcher (PM), MO18 'Nuclear issues', 20 October 1982. TNA PREM 19/979.

MOD, Letter, Pym to Private Secretary to General Synod of the Church of England (General Synod debate on 'The Church and the Bomb'), 24 February 1983. Margaret Thatcher Foundation Archive: TNA PREM 19/1960.

MOD, *British Military Doctrine*, Joint Warfare Publication 0-01 (2001). London: MOD. Available: www.navedu.navy.mi.th/stg/databasestory/data/youttasart/youttasarttalae/bigcity/UK/British%20Defence%20Doctrine.pdf [accessed 12 September 2019].

MINISTRY OF ECONOMIC WARFARE. 'Night bombing as an instrument of economic warfare', 4 February 1942, briefing by OL Lawrence. Reproduced in Webster and Frankland (1961), *The Strategic Air Offensive against Germany 1939–1945*, vol. IV, London, HMSO, pp. 214–19.

PENNEY, W. Unreferenced handwritten notes to Portal on High Explosive Research project 1946–7. TNA AB16/1905.

TRENCHARD, H. Memorandum by the Chief of the Air Staff for the Chiefs of Staff Sub-Committee on the War Object of an Air Force, 2 May 1928. Reproduced in Webster and Frankland (1961), *The Strategic Air Offensive against Germany 1939–1945*, vol. IV, London, HMSO, pp. 71–6.

Other government papers

DEPARTMENT OF EDUCATION AND SCIENCE, Letter to Medical Research Council: '*In vitro* fertilisation: possible enquiry into medical ethics', 13 April 1982. TNA FD7-2307.

MEDICAL RESEARCH COUNCIL, 'Advisory group to review policy on research on *in vitro* fertilisation and embryo transfer in humans: minutes of the meeting held on Tuesday 6 March 1979.' TNA FD7-2307.

MEDICAL RESEARCH COUNCIL, Letter D409/191, 22 November 1982. TNA FD7-2307.

TREASURY, Unreferenced personal memorandum, Polaris successor; memorandum, 16 June 1980. TNA PREM 19/417.

WARNOCK INQUIRY, Letter, 'Research related to human fertilisation and embryology', 25 January 1983. TNA FD7/2307.

Other primary sources

Anderson, S.J. Advisory Committee on Atomic Energy, 'Memorandum on international control of atomic energy', 5 October 1945. TNA CAB 130/3.
Anstey, A. SG(60)35, 'Note on the concept and definitions of breakdown', 10 June 1960. TNA DEFE 10/402.
Blair, T. Blair's resignation speech [online], 2007. London: Guardian Newspapers. Available: www.guardian.co.uk/politics/2007/may/10/labourleadership.labour2 [accessed 15 January 2013].
Browne, D. 'The United Kingdom's nuclear deterrent in the 21st century', 2007. London: Guardian. Available: www.theguardian.com/politics/2007/jan/25/immigrationpolicy.nuclear [accessed 12 September 2019].
Conservative Party, Letter, Jopling (Chief Whip) to Whitelaw (Home Secretary), Unreferenced, 9 May 1980. TNA PREM 19/689.
Conservative Party, Unreferenced memorandum, Ingham (Press Office) to Prime Minister, 'Media relations – stocktaking and looking ahead', 3 August 1982. Margaret Thatcher Foundation Archive. www.margaretthatcher.org.
Conservative Party, Unreferenced memorandum, Ingham (Press Office) to Butler (Cabinet Office), 'Liaison committee', 9 September 1982. Margaret Thatcher Foundation Archive. www.margaretthatcher.org.
Conservative Party, Unreferenced 'Note of Liaison Committee Meeting, Friday 10 Sept[ember] 1982'. Margaret Thatcher Foundation Archive. www.margaretthatcher.org.
Conservative Party, Unreferenced 'Note of Liaison Committee Meeting, Wednesday 20 Oct[ober] 1982'. Margaret Thatcher Foundation Archive. www.margaretthatcher.org.
Conservative Party, Conservative general election manifesto, 1979. London.
Fox, L. 2010. 'Deterrence in the 21st century: speech at Chatham House', 13 July 2010. London: Chatham House. Available: www.chathamhouse.org/sites/default/files/public/Meetings/Meeting%20Transcripts/130710fox.pdf [accessed 12 September 2019].
International Red Cross and Red Crescent Movement. 'International Humanitarian Law Rule 145: Reprisals' [online, 2016]. Available: https://ihl-databases.icrc.org/customary-ihl/eng/print/v1_rul_rule145 [accessed 13 September 2016].

King George VI. 'The King's speech to his peoples' [online]. London: BBC. Available: www.historic-uk.com/HistoryUK/HistoryofBritain/The-Kings-Speech/ [accessed 13 April 2013].
Labour Party, 'The new Britain', manifesto (1964). Available: http://labourmanifesto.com/1964/1964-labour-manifesto.shtml [accessed 23 September 2015].
Labour Party, 'The Labour way is the better way', manifesto (1979). Available: www.politicsresources.net/area/uk/man/lab79.htm [accessed 23 September 2015].
NATO, 'The alliance's new strategic concept', [online] Rome: NATO, 1991. Available: www.nato.int/cps/en/natohq/official_texts_23847.htm [accessed 13 September 2016].
NATO, 'The alliance's strategic concept', Brussels: NATO, 1999. Available: www.nato.int/cps/en/natohq/official_texts_27433.htm [accessed 12 March 2017].
NATO, 'Active engagement, modern defence', Brussels: NATO, 2010. Available: www.nato.int/nato_static_fl2014/assets/pdf/pdf_publications/20120214_strategic-concept-2010-eng.pdf [accessed 23 August 2021].
NATO, 'Warsaw Summit communique', [online], Brussels: NATO, 2016. Available: www.nato.int/cps/en/natohq/official_texts_133169.htm?selectedLocale=en [accessed 13 September 2016], para. 54.
Vance, C. 'Statement to the press after NATO High Level Group special meeting 12 Dec 79', US State Department (1979). Available: http://usa.usembassy.de/etexts/vance5688.htm [accessed 25 October 2015].
Watkins, P. 'Peter Watkins; filmmaker/media critic' [online, 2007]. MNSI. Available: http://pwatkins.mnsi.net/index.htm [accessed 3 January 2013].

Secondary sources

Acheson, R. (ed.) *Assuring destruction forever; nuclear weapon modernization around the world*. New York: Women's International League for Peace and Freedom, 2012.
Air Intelligence Department. *Air defence of Great Britain*, Vol. III. TNA AIR 41/17.
Air Staff, 'Bomber Command, the Air Ministry account of Bomber Command's offensive against the Axis' (London: HMSO; September 1939–July 1941).
Andrew, C. *Defence of the realm: the authorised history of MI5*. Kindle edn. London: Penguin, 2009.
Augustine, St. *City of God*. Chicago: Encyclopaedia Britannica, 1952.
Barnett, C. *The collapse of British power*. Humanities Press Great Britain, 1972.

BASIC. 'Trident Commission concluding report; an independent, cross-party inquiry to examine UK nuclear weapons policy.' London: British American Security Information Council, 2014.
Bartlett, C. *The long retreat: a short history of British defence policy 1945–1970.* London: Macmillan, 1972.
Baylis, J. *Ambiguity and deterrence: British nuclear strategy 1945–1964,* Oxford: Oxford University Press, 1995.
Baylis, J. *Nuclear ethics, realism and utopianism: the 'permanent dialogue' revisited.* Swansea: University of Wales, 2002.
Baylis, J. and Stoddart, K. 'The British nuclear experience: the role of ideas and beliefs (part one)'. *Diplomacy & Statecraft,* 23 (2012), 331–46.
Baylis, J. and Stoddart, K. 'The British nuclear experience: the role of beliefs, culture, and status (part two)'. *Diplomacy & Statecraft,* 23 (2012), 493–516.
Baylis, J. and Stoddart, K. *The British nuclear experience: the roles of beliefs, culture and identity.* Oxford: Oxford University Press, 2014.
Bellamy, A. *Just wars: from Cicero to Iraq.* London: Blackwell, 2006.
Bellamy, A. 'The ethics of terror bombing: beyond supreme emergency'. *Journal of Military Ethics,* 7 (2008), 41–65.
Bellamy, A. 'When is it right to fight? International law and *jus ad bellum*'. *Journal of Military Ethics,* 8 (2009), 231–45.
Benn, T. *Parliament, people and power; agenda for a free society.* London: Verso, 1982.
Benn, T. *The Benn diaries: 1940–1990.* Kindle edn. Harmondsworth: Arrow Books, 2013.
Bevan, A. 'Labour Party Conference speech, Brighton, 3 October 1957' in Brian MacArthur, *The Penguin Book of Twentieth-Century Speeches* (London: Penguin, 1994).
Biddle, T. D. *Rhetoric and reality in air warfare: the evolution of British and American ideas about strategic bombing 1914–1945.* Kindle edn. Princeton, NJ: Princeton University Press, 2002.
Booth, K. and Barnaby, F. *The future of Britain's nuclear weapons: experts reframe the debate.* Oxford: Oxford Research Group, 2006.
Bouwman, B. 'Present at the undoing: the Netherlands and the multilateral force'. Nuclear Proliferation International History Project [online], 2013. Available: www.wilsoncenter.org/publication/present-the-undoing-the-netherlands-and-the-multilateral-force [accessed 27 July 2017].
Briggle, A. and Mitcham, C. *Ethics and science: an introduction.* Cambridge: Cambridge University Press, 2012.
Brittain, V. 'Seed of chaos: what mass bombing really means.' London: New Vision Publishing Co., 1944.
Bronson, R. 'It is two and a half minutes to midnight; 2017 Doomsday Clock Statement'. Bulletin of the Atomic Scientists [online]. Chicago:

Bulletin of the Atomic Scientists, 2017. Available: http://thebulletin.org/sites/default/files/Final2017ClockStatement.pdf [accessed 27 January 2017].

Browne, D. *The United Kingdom's nuclear deterrent in the 21st century*. London: The Stationery Office, 2007.

Browne, S. S. S. 'Right acts and moral actions'. *The Journal of Philosophy*, 42 (1945), 505–15.

Bundy, M., Kennan, G. F., McNamara, R. and Smith, G. 'Nuclear weapons and the Atlantic alliance'. *Foreign Affairs*, 60 (1982).

Burke, D. 'GM food and crops: what went wrong in the UK?' *EMBO Reports*, 5 (2004), 432–6.

Burke, E. 'Speech to the electors of Bristol' [online]. Chicago: The University of Chicago Press, (1774; London: Henry G. Bohn, 1854–6; 1986). Available: https://press-pubs.uchicago.edu/founders/documents/v1ch13s7.html [accessed 22 July 2011].

Butler, L. and Gorst, A. (eds). *Modern British history; a guide to study and research*. London: I. B. Tauris.

Butterfield, H. *History and human relations*. London: Collins, 1951.

Byrne, P. 'Nuclear weapons and CND'. *Parliamentary Affairs*, 51 (1998), 424–34.

Carr, E. H. *The twenty years' crisis 1919–1939: an introduction to the study of international relations*. London: Macmillan and Co., 1962.

Cathcart, B. *Test of greatness: Britain's struggle for the atomic bomb*. Kindle edn. UK: Endeavour Press, 1994.

CBS News, 'Remembering Cuban missile crisis, 50 years later' [online]. Washington DC: CBS, 2012. Available: www.cbsnews.com/news/remembering-cuban-missile-crisis-50-years-later/ [accessed 17 December 2012].

Chapman, J. 'The BBC and the censorship of *The War Game* (1965)'. *Journal of Contemporary History*, 41 (2006), 75–94.

Chapman, J. '*The War Game* controversy – again'. *Journal of Contemporary History*, 43 (2008), 8.

Chickering, R. and Forster, S. (eds) *A world at total war: global conflict and the politics of destruction*. Cambridge: Cambridge University Press, 2005.

Chifor, G. 'War and self-defence by David Rodin'. *The Modern Law Review*, 66 (2003), 483–8.

Childress, J. F. '*Just and Unjust War*: review'. *The Bulletin of the Atomic Scientists* (1978).

Childress, J. F. 'Just war theories: the bases, interrelations, priorities and functions of their criteria'. *Theological Studies*, 39 (1978), 427–45.

Childress, J. F. 'Can modern war be just?' *Journal of the American Academy of Religion*, 55 (1987), 604.

Church of Scotland. 'Nuclear weapons: society, religion and technology project'. Edinburgh: Church and Society Council of the Church of Scotland, 2014. [online] Available: www.churchofscotland.org.uk/data/assets/pdf_file/0003/8895/Nuclear_Weapons_leaflet.pdf [accessed 12 September 2019].
Church, W. F. *Richelieu and reasons of state*. Princeton, NJ: Princeton University Press, 1972.
Churchill, W. *The gathering storm*. New York: Rosetta Stone, 2009.
Churchill, W. *Their finest hour*. New York: Rosetta Stone, 2010.
Churchman, C. W. 'Free for all'. *Management Science*, 14 (1967), B141–6.
Clausewitz, C. von *On war*, Howard, M. and Paret, P. (eds). Kindle edn. Princeton, NJ: Princeton University Press, 2008.
CND. 'CND free teaching resources', [online]. London: CND, 2016a. Available: www.cnduk.org/information/peace-education/teaching-resources [accessed 22 November 2016].
CND. 'CND Trident mythbuster', [online]. London: CND, 2016b. Available: www.cnduk.org/images/stories/Trident_mythbuster_2016.pdf [accessed 22 November 2016].
CND. 'No to Trident', [online]. London: CND, 2016c. Available: www.cnduk.org/campaigns/no-to-trident [accessed 22 November 2016].
Cochrane, A. and Anderson, J. (eds) *Restructuring Britain: politics in transition*. London: Sage.
Coker, C. *Humane warfare*. London: Routledge, 2001.
Coleman, S. 'Reconsidering the supreme emergency argument'. 'When is it right to do wrong?' Unpublished paper presented at European International Society for Military Ethics Conference, 2012, Shrivenham, UK. ISME.
Connelly, M. 'The British people, the press and the strategic air campaign against Germany, 1939–45'. *Contemporary British History*, 16 (2002), 2, 39–48.
Cook, R. *The point of departure*. London: Simon and Schuster, 2003.
Corbett, A. 'Deterring nuclear Russia in the 21st century; theory and practice'. NATO Defense College Research Report. Rome: NATO Defense College, 2016.
Corbett, A. S. '*Igitur qui desiderat pacem, praeparet bellum atomica*'. *Journal of Military Ethics*, 19:4 (2020), 331–47.
Cox, B. and Forshaw, J. 'Education is as important to security as aircraft carriers or missile defence'. *The Big Issue*, 21 November 2016.
Crossman, R. *The backbench diaries of Richard Crossman*. London: Hamish Hamilton and Jonathan Cape, 1981.
Darwin, J. *Britain and decolonisation: the retreat from empire in the post war world*. London: Macmillan, 1988.

Darwin, J. *The end of the British Empire*. Oxford: Blackwell, 1991.
Day, R. *... but with respect: memorable interviews with statesmen and Parliamentarians*. London: Weidenfeld and Nicolson, 1993.
Dessler, D. 'Beyond correlations: toward a causal theory of war'. *International Studies Quarterly*, 35 (1991), 337–55.
Donnelly, S. 'CND: the story of a peace movement'. *History Today*, 58 (2008).
Dorril, S. *The silent conspiracy*. London: Heinemann, 1993.
Douhet, G. *The command of the air*. Washington, DC: Air Force History and Museum Programme, (1921; 1998).
Fallon, M. 'The case for the retention of the UK's independent nuclear deterrent', speech at the Policy Exchange, 23 March 2016 [online]. London: HMSO, 2016. Available: www.gov.uk/government/speeches/the-case-for-the-retention-of-the-uks-independent-nuclear-deterrent [accessed 22 November 2016].
Fisher, D. *Morality and the bomb; an ethical assessment of nuclear deterrence*. Beckenham: Croom Helm, 1985.
Fitch, R. E. 'An experimental Christian ethics'. *The Journal of Religion*, 20 (1940), 325–39.
Folwell, A. 'Opinion formers poll: views on Trident highly politicised as Scotland vote approaches'. Politics, Reputation Research, Scotland, UK. London: YouGov, 2014.
Frankland, N. *The bombing offensive against Germany: outlines and perspectives*. London: Faber and Faber, 1965.
Freedman, L. *Britain and nuclear weapons*. London: Macmillan, 1980.
Freedman, L. 'Britain: the first ex-nuclear power?' *International Security*, 6 (1981), 80–104.
Freedman, L. *The evolution of nuclear strategy*. New York: St Martin's Press, 1981.
Freedman, L. 'NATO myths'. *Foreign Policy* (1981), 48–68.
Freedman, L. *Deterrence*. Cambridge: Polity Press, 2004.
Frewer, L. J., Howard, C. and Shepherd, R. 'Public concerns in the United Kingdom about general and specific applications of genetic engineering: risk, benefit, and ethics'. *Science, Technology, & Human Values*, 22 (1997), 98–124.
George, B. 'Political perspectives on the outcome of SDR: the House of Commons Select Committee report'. *The RUSI Journal*, 143 (1998), 26.
Gill, D. J. 'Nuclear deterrence and the tradition of non-use'. *International Affairs*, 85 (2009), 863–9.
Goldberg, A. 'The atomic origins of the British nuclear deterrent'. *International Affairs (Royal Institute of International Affairs 1944–)*, 40 (1964), 409–29.

Goldberg, A. 'The military origins of the British nuclear deterrent'. *International Affairs (Royal Institute of International Affairs 1944–)*, 40 (1964), 600–618.
Goldhamer, J. 'The economist's perception of the US–Soviet strategic balance: an update for 1979–1981'. Santa Monica, CA: RAND Corporation, 1983.
Gowing, M. *Independence and deterrence: Britain and atomic energy 1945–1952*, vol. 1. London: Macmillan, 1974.
Gowing, M. *How nuclear power began*. Southampton University, 1987.
Grant, M. *After the bomb: civil defence and nuclear war in Britain 1945–68* (London: Palgrave Macmillan, 2009).
Gray, C. S. and Howard, M. 'Perspectives on fighting nuclear war'. *International Security*, 6 (1981), 185–7.
Gray, C. S. *Another bloody century*. London: Orion, 2005.
Gray, P. 'The gloves will have to come off: a reappraisal of the legitimacy of the RAF bomber offensive against Germany'. *Air Power Review*, 13 (2010), 9–40.
Gray, P. *The leadership, direction and legitimacy of the RAF bomber offensive from inception to 1945*. London: Bloomsbury Academic, 2012.
Grayling, A. C. *Among the dead cities*. London: Bloomsbury, 2006.
Greenpeace. 'Trident – the UK's nuclear weapons system' [online]. Greenpeace, 2006. Available: www.greenpeace.org.uk/peace/trident-the-uks-nuclear-weapons-system [accessed 15 July 2016].
Gregg, V. *Dresden: a survivor's story*. London: Bloomsbury Reader, 2013.
Groom, A. *British thinking about nuclear weapons*. London: Pinter Publishers, 1974.
Guthrie, C. and Quinlan, M. *Just war. The just war tradition: ethics in modern warfare*. London: Bloomsbury, 2007.
Haidt, J. *The righteous mind: why good people are divided by politics and religion*. Kindle edn. London: Allen Lane, 2012.
Haines, S. 'Replacing Trident: a new nuclear debate?' *Naval Review*, 93 (2005), 113–17.
Hardin, R. and Freedman, L. *Nuclear deterrence: ethics and strategy*. Chicago/ London: University of Chicago Press, 1985.
Harris, A. *The bomber offensive*. Barnsley: Collins, 1947.
Harrison, M. L. 'CND: the challenge of the post-Cold-War era'. PhD thesis, University of Loughborough, 1994.
Haslam, J. *No virtue like necessity: realist thought in international relations since Machiavelli*. New Haven, CT and Oxford: Yale University Press, 2002.
Hastings, M. *Bomber Command*. Basingstoke and Oxford: Pan, 2010.
Healey, D. *The time of my life*. London: Michael Joseph, 1989.

Hennessy, P. *Muddling through: power, politics and the quality of government in postwar Britain*. Oxford: Indigo, 1996.
Hennessy, P. *Having it so good: Britain in the Fifties*. London: Penguin Books, 2006.
Hennessy, P. 'Cabinets and the Bomb' workshop [online]. London: British Academy, 2007. Available: www.britac.ac.uk/node/4986/ [accessed 27 July 2017].
Hennessy, P. *The secret state*. London: Penguin Books, 2010.
Hennessy, P. and Jinks, J. *The silent deep: the Royal Navy submarine service since 1945*. UK: Penguin Random House, 2015.
Hipperson, S. 'Greenham Common women's peace camp' (online: 2000). Available: www.greenhamwpc.org.uk [accessed 17 April 2016].
Hobsbawm, E. *Age of extremes: the short twentieth century*. London: Abacus, 1995.
Hogg, J. *British nuclear culture: official and unofficial narratives in the long 20th century*. London: Bloomsbury, 2016.
Holman, B. 'Airminded – airpower and British society 1908–1941' [online: 2006]. University of New England, Armidale, Australia. Available: http://airminded.org/biographies/j-m-spaight/ [accessed 16 March 2014].
Hudson, K. *CND now more than ever: the story of a peace movement*. Kindle edn. Matrix Digital Publishing, 2007.
Irish Bishops' Conference on War and Peace in the Nuclear Age. 'The storm that threatens: statement by the Irish Bishops' Conference on War and Peace in the Nuclear Age'. *The Furrow*, 34 (1983), 589–95.
Inglehart, R. 'Post-materialism in an environment of insecurity'. *American Political Science Review*, 75 (1981), 880–99.
John XXIII, Pope, '*Pacem in terris*: Encyclical of Pope John XXIII, 11 April 1963'. Available: www.vatican.va/content/john-xxiii/en/encyclicals/documents/hf_j-xxiii_enc_11041963_pacem.html [accessed 17 October 2019].
Johnson, J. T. 'Toward reconstructing the *jus ad bellum*'. *The Monist*, 57 (1973).
Johnson, J. T. *Can modern war be just?* New Haven, CT: Yale University Press, 1984.
Johnson, J. T. *Morality and contemporary warfare*. New Haven, CT: Yale University Press, 1999.
Johnson, J. T. 'The idea of defense in historical and contemporary thinking about just war'. *The Journal of Religious Ethics*, 36 (2008), 543–56.
Johnson, J. T. 'Thinking historically about just war'. *Journal of Military Ethics*, 8 (2009), 246–59.
Johnson, J. T. *Ethics and use of force: just war in historical perspective*. London: Ashgate, 2011.

Jones, H. A. *The war in the air*, Appendices: 'Being the story of the part played in the Great War by the Royal Air Force', London: Naval & Military Press, 2002.
Jones, N. *The origins of strategic bombing: a study of the development of British air strategic thought and practice up to 1918*. London: Harper Collins, 1973.
Jourdain, M. 'Air raid reprisals and starvation by blockade'. *International Journal of Ethics*, 28 (1918), 542–53.
Kahn, H. *On thermonuclear war*. Princeton, NJ: Princeton University Press, 1961.
Kahn, H. *On escalation: metaphors and scenarios*. New York: Frederic A. Praeger, 1969.
Kaplan, F. *The wizards of Armageddon*. Stanford, CA: Stanford University Press, 1991.
Kaplan, M. A. 'The calculus of nuclear deterrence'. *World Politics*, 11 (1958), 20–43.
Kaufmann, G. 'Thatcher triumphs again' [online]. London: BBC, 1983. Available: http://news.bbc.co.uk/2/hi/uk_news/politics/vote_2005/basics/4393313.stm#issues [accessed 25 October 2015].
Kellner, P. 'Do we want Trident? Most people want to protect defence spending – even at the expense of other services'. *Prospect*, August 2015.
Kenny, A. *The logic of deterrence: a philosopher looks at the arguments for and against nuclear disarmament*. London: Waterstone & Co. Firethorne Press, 1985.
Kent, B. 'Protest and survive'. *History Today*, 49 (1999), 14–16.
Kim, S. S. 'The US Catholic bishops and the nuclear crisis'. *Journal of Peace Research*, 22 (1985), 321–33.
Lancet, The: Editorial. 'Traffic in dead bodies'. *The Lancet* (1828), 818–21.
Laucht, C. *Elemental Germans: Klaus Fuchs, Rudolf Peierls and the making of British nuclear culture 1939–59*. London: Palgrave Macmillan, 2012.
Lawrence, L. *Children of the dust*. London: Random House, 1985.
Lee, G. 'I see dead people: air-raid phobia and Britain's behaviour in the Munich crisis'. *Security Studies*, 13 (2010), 42.
Leira, H. 'Justus Lipsius, political humanism and the disciplining of 17th-century statecraft'. *Review of International Studies*, 34 (2008), 669–92.
Letwin, S. R. 'Hobbes and Christianity'. *Daedalus*, 105 (1976), 1–21.
Lowe, K. *Inferno: the devastation of Hamburg 1943*. London: Penguin Group, 2007.
Maier, C. S. 'Targeting the city: debates and silences about the aerial bombing of World War II'. *International Review of the Red Cross*, 87 (2005), 429–44.
Machiavelli, N. *The Prince*. Cambridge: Cambridge University Press, (1513; 1988).

Macmillan, H. and Catterall, P. *The Macmillan diaries*, Vol. II, *Prime Minister and after*. London, Pan Macmillan, 2011.
Macmillan, H. and Catterall, P. *The Macmillan diaries*, Vol. I, *The Cabinet years 1950–1957*. London: Pan, 2012.
Maret, S. (ed.) *Government secrecy*. Bingley, UK: Emerald Books.
Markus, G. 'Comment'. *New Blackfriars*, 72 (1991), 54–5.
Marr, A. *My trade*. London: Pan Macmillan, 2004.
Marsh, C. and Fraser, C. (eds) *Public opinion and nuclear weapons*. Basingstoke: Macmillan Press, 1989.
McMahan, J. 'Deterrence and deontology'. *Ethics*, 95 (1985), 517–36.
McMahan, J. 'Is nuclear deterrence paradoxical?' *Ethics*, 99 (1989), 407–22.
McMahan, J. 'Just cause for war'. *Ethics & International Affairs* (Wiley-Blackwell), 19 (2005), 1–21.
McMahan, J. *Killing in war*. Oxford: Oxford University Press, 2009.
Mearsheimer, J. 'E. H. Carr vs idealism: the battle rages on'. *International Relations*, 19 (2005), 139–52.
Mercer, P. *'Peace' of the dead: the truth behind the nuclear disarmers*. London: Policy Research Publications, 1986.
Middlebrook, M. and Everitt, C. *The Bomber Command war diaries: an operational reference book 1939–1945*. Harmondsworth: Viking Penguin, 1985.
Miller, L. H. 'The contemporary significance of the doctrine of just war'. *World Politics*, 16 (1964), 254–85.
Minnion, J. and Bolsover, P. *The CND story: the first 25 years of CND in the words of the people involved*. London: Allison and Busby, 1983.
Morgan, K. *Callaghan: a life*. New York: Oxford University Press, 1997.
Morgenthau, H. J. 'The evil of politics and the ethics of evil'. *Ethics*, 56 (1945), 1–18.
Nott, J. *Here today, gone tomorrow*. London: Politico's, 2002.
O'Brien (Cardinal). 'Nuclear weapons – replacing Trident – a Scottish Catholic response'. In: Archdiocese of St Andrews and Edinburgh (ed.) 'Justice and peace pamphlet'. Glasgow: Justice and Peace Scotland, 2006.
O'Connor, J. *The meaning of crisis: a theoretical introduction*. Oxford: Blackwell, 1987.
Ogilvie-White, T. *On nuclear deterrence: the correspondence of Sir Michael Quinlan*. London: Routledge, 2011.
Olry, R. 'Body snatchers: the hidden side of the history of anatomy'. *Journal of the International Society of Plastination*, 14 (1999), 6–9.
Orwell, G. *Nineteen eighty four*. Harmondsworth: Penguin, 1949.
Overpeck, D. '"Remember! it's only a movie!" Expectations and receptions of *The Day After* (1983)'. *Historical Journal of Film, Radio and Television*, 32 (2012), 267–92.

Overy, R. J. 'Hitler and air strategy'. *Journal of Contemporary History*, 15 (1980), 405–21.
Overy, R. J. 'Air power and the origins of deterrence theory before 1939'. *Journal of Strategic Studies*, 15 (1992), 28.
Overy, R. J. *Why the Allies won*. London: W. W. Norton, 1997.
Overy, R. J. *The bombing war: Europe 1939–1945*. London: Penguin, 2013.
Owen, D. *The politics of defence*. London: Jonathan Cape, 1972.
Owen, D. *Nuclear papers*. Liverpool: Liverpool University Press, 2009.
Pearson, R. 'More than would be reasonably anticipated: No. 3 Wing Royal Naval Air Service' (2013). Available: www.overthefront.com/over-the-front-journal/sample-articles/more-than-would-be-reasonably-anticipated [accessed 15 December 2013].
Pielke, R. A. Jr, 'When scientists politicize science: making sense of controversy over *The skeptical environmentalist*'. *Environmental Science & Policy*, 7 (2004), 405–17.
Pincher, C. *Into the atomic age*. London: Hutchinson & Co., 1948.
Plato. *The Republic*. Chicago: Encyclopedia Britannica, 1952.
Preston, D. *A higher form of killing: six weeks in World War One that forever changed the nature of warfare*. Kindle edn. London and New York: Bloomsbury, 2015.
Preston, P. *The destruction of Guernica*. London: Harper Press, 2012.
Quade, Q. L. 'Civil disobedience and the state'. *Worldview*, 10:11 (1967), 4–9.
Quinlan, M. *Thinking about nuclear weapons: principles, problems, prospects*. Oxford: Oxford University Press, 2009.
Quinlan, M. 'Thinking about nuclear weapons'. *The RUSI Journal*, 142 (1997), 1.
Reading, S. 'Why film should not be shown', *The Times* (1 March 1966), 13.
Renfrew, C. *Prehistory: the making of the human mind*. London: Phoenix EBooks, 2007.
Ritchie, N. *Trident: the deal isn't done. Serious questions remain unanswered*. Bradford: Bradford University, 2007.
Ritchie, N. 'Deterrence dogma? Challenging the relevance of British nuclear weapons'. *International Affairs*, 85 (2009), 81–98.
Ritchie, N. 'Relinquishing nuclear weapons: identities, networks and the British bomb'. *International Affairs (Royal Institute of International Affairs 1944–)*, 86 (2010), 465–87.
Ritchie, N. and Ingram, P. 'A progressive nuclear policy: rethinking continuous at-sea deterrence'. *RUSI Journal*, 155 (2010).
Ritchie, N. *A nuclear weapons-free world?: Britain, Trident and the challenges ahead*. London: Palgrave Macmillan, 2012.

Riviere, P. 'Unscrambling parenthood: the Warnock Report'. *Anthropology Today*, 1 (1985), 5.
Roberts, B. *The case for US nuclear weapons in the 21st century*. Stanford, CA: Stanford University Press, 2016.
Robinson, E., Schofield, C., Sutcliffe-Braithwaite, F. and Thomlinson, N. 'Telling stories about post-war Britain: popular individualism and the "crisis" of the 1970s'. *Twentieth Century British History*, 28 (2017), 268–304.
Rodin, D. *War and self-defense*. Oxford: Oxford University Press, 2002.
Rodin, D. and Shue, H. (eds) *Just and unjust warriors: the moral and legal status of soldiers*. Oxford: Oxford University Press, 2008.
Rodin, D. 'Justifying harm'. *Ethics*, 122 (2011), 74–110.
Rosenthal, A. *New documentary in action: casebook in film making*. Berkeley: University of California Press, 1971.
Ruddock, J. *Going nowhere: a memoir*. Kindle edn. London: Biteback Publishing, 2016.
Royal United Services Institute, 'The future of the UK strategic deterrent', London: RUSI, 2005.
Sabin, P. A. *The Third World War scare in Britain*. Basingstoke: Macmillan Press, 1986.
Sandman, P. 'Explaining risk to non-experts: a communications challenge'. *Emergency Preparedness Digest* (1987a), 25–9.
Sandman, P. 'Risk communication: facing public outrage'. *EPA Journal* (US Environmental Protection Agency) (1987b), 21–2.
Sandman, P. and Valenti, J. 'Scared stiff – or scared into action'. *Bulletin of the Atomic Scientists* (1986), 12–16.
Schafer, M. and Crichlow, S. 'The "process–outcome" connection in foreign policy decision making: a quantitative study building on groupthink'. *International Studies Quarterly*, 46 (2002), 45–68.
Schelling, T. C. *Arms and influence*. Kindle edn. New Haven, CT and London: Yale University Press, 2008.
Schlosser, E. *Command and control*. Kindle edn. London: Penguin, 2014.
Scott, L. 'Labour and the bomb: the first 80 years'. *International Affairs*, 82 (2006), 685–700.
Sharma, S. K. 'The legacy of *jus contra bellum*: echoes of pacifism in contemporary just war thought'. *Journal of Military Ethics*, 8 (2009), 217–30.
Sharp, P. 'From yellow peril to Japanese wasteland: John Hersey's *Hiroshima*'. *Twentieth Century Literature*, 46 (2000), 434–52.
Shaw, T. 'The BBC, the state and cold war culture: the case of television's *The War Game* (1965)'. *The English Historical Review*, 121 (2006), 34.
Shea, J. '1979: the Soviet Union deploys its SS20 missiles and NATO responds', NATO lecture (2009). Available: www.nato.int/cps/en/natohq/opinions_139274.htm [accessed July 2013].

Sherman, N. *Stoic warriors: the ancient philosophy behind the military mind*. New York: Oxford University Press, 2005.
Shue, H. and Freedman, L. *Nuclear deterrence and moral restraint: critical choices for American strategy*. Cambridge: Cambridge University Press, 1989.
Siemes, P. T. 'The atomic bomb on Hiroshima: an eye-witness account'. *The Irish Monthly*, 74 (1946), 93–104.
Siemes, P. T. 'The atomic bomb on Hiroshima: an eye-witness account (continued)'. *The Irish Monthly*, 74 (1946), 148–54.
Singer, J. 'The correlates of war project'. *World Politics*, 24 (1972), 243–70.
Slim, H. *Killing civilians: method, madness and morality in war*. London: Hurst, 2007.
Spaight, J. 'The war in the air; first phase'. *Foreign Affairs*, 18 (1940).
Spaight, J. 'The war in the air; second phase'. *Foreign Affairs*, 19 (1941), 408–13.
Spaight, J. M. *Air power in the next war*. London: Geoffrey Bles, 1938.
Spaight, J. M. *Bombing vindicated*. London: Geoffrey Bles, 1944.
Starkswood, D. 'Do we need to replace Trident?' [online, 2016]. Available: www.facebook.com/ThatDamnedIdealistAgain?lst=100000442255266%3A606372421%3A1483013236 [accessed 29 December 2016].
Stocker, J. 'The United Kingdom and nuclear deterrence'. Adelphi Papers, 2007.
Stoddart, K. 'Maintaining the "Moscow criterion": British strategic nuclear targeting 1974–1979'. *Journal of Strategic Studies*, 31 (2008), 897–924.
Stoddart, K. *Losing an empire and finding a role: Britain, the USA, NATO and nuclear weapons, 1964–70*. London: Palgrave Macmillan, 2012.
Stoddart, K. *The sword and the shield: Britain, America, NATO and nuclear weapons, 1970–1976*. London: Palgrave MacMillan, 2014.
Stoddart, K. *Facing down the Soviet Union: Britain, the USA, NATO and nuclear weapons 1976–1983*. London: Palgrave Macmillan, 2014.
Swindells, R. *Brother in the land*. London: Oxford University Press, 1984.
Taylor, F. *Dresden: Tuesday 13 February 1945*. London: Bloomsbury, 2011.
Taylor, R. and Pritchard, C. *The protest makers: the British nuclear disarmament movement of 1958–1965, twenty years on*. Oxford: Pergamon, 1980.
Tertrais, B. 'In defense of deterrence: the relevance, morality and cost-effectiveness of nuclear weapons', Proliferation Papers (2011). Available: www.ifri.org/en/publications/enotes/proliferation-papers/defense-deterrence-relevance-morality-and-cost [accessed 23 November 2016].
US National Conference of Catholic Bishops, The. 'The challenge of peace: God's promise and our response – a pastoral letter on war and peace'.

Origins (1983): 2. Available: www.usccb.org/upload/challenge-peace-gods-promise-our-response-1983.pdf [accessed 17 October 2019].

Van Voorst, L. B. 'The churches and nuclear deterrence'. *Foreign Affairs*, 61 (1983), 827–52.

Walzer, M. 'Political action: the problem of dirty hands'. *Philosophy and Public Affairs*, 2 (1973), 160–80.

Walzer, M. *Just and unjust wars: a moral argument with historical illustrations*. New York: Basic Books, 1977.

Warnock, M. 'Moral thinking and government policy: the Warnock Committee on Human Embryology'. *The Milbank Memorial Fund Quarterly; Health and Society*, 63 (1985), 19.

Wayne, M. 'Failing the public: the BBC, *The War Game* and revisionist history: a reply to James Chapman'. *Journal of Contemporary History*, 42 (2007), 11.

Webster, C. and Frankland, N. *The strategic air offensive against Germany 1939–1945*, Vol. I, *Preparation*. London: HMSO, 1961.

Webster, C. and Frankland, N. *The strategic air offensive against Germany 1939–1945*, Vol. III, *Victory*. London: HMSO, 1961.

Webster, C. and Frankland, N. *The strategic air offensive against Germany 1939–1945*, Vol. IV, Appendices. London: HMSO, 1961.

Wells, H. G. *The war in the air, and particularly how Mr. Bert Smallways fared while it lasted*. London: George Bell and Sons, 1908.

Wells, H. G. *The world set free: a story of mankind*. London: Macmillan and Co., 1914.

Whetham, D. '"Are we fighting yet?" Can traditional just war concepts cope with contemporary conflict and the changing character of war?' *The Monist*, 99 (2016), 55–69.

Wilkinson, N. *Secrecy and the media: the official history of the United Kingdom's D-notice system*. Abingdon: Routledge, 2009.

Willett, L. 'Renewing Britain's independent strategic nuclear deterrent: a debate' [online]. RUSI, 2007. Available: https://rusi.org/commentary/renewing-britain's-independent-strategic-nuclear-deterrent-debate [accessed 20 November 2016].

Williams, F. *Twilight of empire: memoirs of Prime Minister Clement Attlee*. New York: A. S. Barnes & Co., 1962.

Williams, P. *Hugh Gaitskell: a political biography*. London: Jonathan Cape, 1979.

Wilson, E. O. *The social conquest of Earth*. Kindle edn. London: W. W. Norton, 2012.

Wilson, W. 'The myth of nuclear deterrence'. *Nonproliferation Review*, 15 (2008), 421–39.

Wilson, W. *Five myths about nuclear weapons*. Boston, MA: Houghton Mifflin Harcourt, 2013.

Woodard, C. 'War and self-defence by David Rodin'. *Mind*, 114 (2005), 453–57.
Young, K. 'The Skybolt crisis of 1962: muddle or mischief?' *Journal of Strategic Studies*, 27 (2004), 614–35.
Zuckerman, S. 'Judgement and control in modern warfare'. *Foreign Affairs*, 40 (1962), 196–212.
Zuckerman, S. *Scientists and war: the impact of science on military and civil affairs*. London: Hamish Hamilton, 1966.

Index

Aaron, D., US deputy National Security Advisor 103, 105
Advisory Committee on Atomic Energy (ACAE) 57
Agriculture and Environment Biotechnology Commission (AEBC) 167–70
Alconbury (RAF station) 120
Aldermaston marches 72–4, 81, 118
Alexander A. V., MP 40, 60, 61
ambiguity 185–6, 226
American Catholic Bishops 126, 152
see also pastoral
anti-ballistic missile defences 85, 87, 98–9, 134
Archbishop of Canterbury 27, 44
arms control 72, 99, 105, 112–13, 131
Armstrong, R., Cabinet Secretary 100–5, 113–14, 199, 210
Asturias 20
Attlee C., Prime Minister 13, 39, 50, 55–60, 62, 66, 101, 110, 181, 191
Augustine of Hippo 18, 148–9, 154, 203

Baker J., US Senator 110
Baldwin S., Prime Minister 28, 50, 193
Balfour H., Under-Secretary for Air 46
BASIC 196, 209–11

Battle of Britain 38
BBC 2, 5, 9–15, 61–2, 73, 123, 215
Benn T., MP 13
Berlin 31, 43, 50, 74, 78, 80
Bevan A., MP 71–3
Blitz 39
Blue Danube 64, 75
Blue Steel 75
Body-snatchers *see* Burke
Bomber Command, targeting 39–41, 43, 47, 48, 51, 146, 194, 202
breakdown 13, 15, 25, 27, 29, 38, 50, 70, 124
Bridges Inquiry 134
British Nuclear Deterrent Study Group 75–6
Brother in the land 124
Brown L. 161
Brzezinsky Z., US National Security Advisor 107
Burke W. 165
Butler R. A., Home Secretary 42, 73, 213

Callaghan J., Prime Minister 74, 86–9, 98, 183
Cameron D., Prime Minister 208
Cameron N., Chief of Defence Staff 109
Campaign for Nuclear Disarmament (CND) 15, 71–74, 78, 81–2, 85, 99, 106, 109, 116–24, 130–6, 152, 186, 191, 198, 205, 212–15

Index

Canton 29
Carrington P., Foreign Secretary 109–10
Carter J., US President 89, 98, 103, 107
Castle Bravo test 65
Central Treaty Organisation (CENTO) 84
Chamberlain N., Prime minister 29–30, 36, 39
Cherwell (Lord) Lindeman F., scientific advisor to Churchill 45
Chevaline 85, 99–01, 108, 112, 135, 182
  see also Polaris
Chiefs of Staff (COS)
  bombing policy 26, 30–1, 40, 45
  nuclear policy 56, 66–8, 76, 81, 107, 199, 202
Chilcot Inquiry 204
Children of the dust 124
Chungking 29
Churchill W., Prime Minister
  bombing policy 3, 24, 38–9, 45, 194, 213
  nuclear policy 13, 63–68, 69, 110, 135, 203
  supreme emergency 3, 135, 145, 153, 203
  civil defence 7–17, 120–3, 136, 189, 200–1
  see also Protect and Survive
Clyde Submarine Base 76–7, 120, 129
  see also Faslane
Cockcroft J. 59
Committee for the Abolition of Night Bombing 44, 213
Committee of 100 12, 74
Committee on Defence Policy 38–40, 51, 58, 68–71, 121, 199
Commons Defence Committee 114–6, 191, 206
competent authority 145
continuous at-sea deterrence 85, 157, 185–9, 210, 212

Cooper F., permanent under-secretary MOD 156
Coronavirus 70, 147, 158, 193, 227
Cost 75, 87, 100, 102, 111, 113, 128, 134–5, 196–8, 210, 224, 226
Crossman R., MP 65, 74–5
Cuban Missile Crisis 78–81, 120

Davies R., MP 49
Day After, The 123, 201, 213
Defence Overseas Policy Committee (DOPC) 121
Defence Secretariat 17 (DS17) 131
Defence Secretariat 19 (DS19) 131–3
Defence Select Committee 88, 108, 111, 226
Department for the Environment, Farming and Rural Affairs (DEFRA) 170
dirty hands 146, 154–6, 203, 225
discrimination 184
D-notice 13, 59–61, 64
double effect 27, 39, 46, 152
Douhet G. 24, 29, 50
Dreadnought 189–92, 197–8, 208, 210, 212, 226
  see also successor
Dresden 50
Duff A. *see* Duff-Mason report
Duff-Mason Report 87, 98, 100, 112, 114, 117, 160

Eden A., Prime Minister 40
embryology 4, 160, 162–3, 167, 180, 226
Enquiry into Human Fertilisation 4, 162–69, 170, 204–5
  see also Warnock M.
extended deterrence 184, 187

Falklands War 129, 162
Fallon M., Secretary of State for Defence 144, 146, 181–2
Faslane 76–7, 120, 129
  see also Clyde Submarine Base

first strike 186–7
Fitch R. 203
Foot M., Leader of the Labour Party 74, 115, 119
Fox L., Secretary of State for Defence 200
Freiburg 20, 23
future United Kingdom strategic nuclear deterrent force, The 112–3, 160, 178

Gaitskell H., Leader of the Labour Party 71, 73–75, 80
GEN 163 Cabinet sub-committee 59
GEN 465 Cabinet sub-committee 64, 69
GEN 75 Cabinet sub-committee 57–9
genetically modified crops 160, 167–71, 177, 180, 226
GM Dialogue 169–70
Gorbachev M., Soviet Premier 136
Gordievsky O., spy 133
Gotha 21, 193
Greenham Common 118, 120, 133
ground-launched cruise missile (GLCM) 99, 118–20, 136, 197
see also intermediate-range nuclear forces
Guernica 29, 50, 55–6, 193
Guthrie C., Chief of Defence Staff 150, 156

Hague conference 1923 25, 31, 36
Hague convention 1907 25
Halifax (Lord) Wood E., Foreign Secretary 37, 55
Hamburg 50
Hard Rock civil defence exercise 122
Hare W. see Burke
Harris A., Commander-in-Chief Bomber Command 39, 40, 47–50
Healey D., Chancellor/Secretary of State for Defence 83, 87, 194, 198–9, 217n.28

Hersey J. 61–2
High Level Group (HLG) 105
see also NATO; Nuclear Planning Group
Hinton C. 59
Hiroshima 18–19, 55, 61–2
Holy Loch 76, 83
Howe G., Chancellor 109–10
Human Fertilisation and Embryology Act 4, 167
Human Fertilisation and Embryology Authority 167
Hunt J., Cabinet Secretary 99–101

Ingham B., Conservative Party press secretary 129
Initial Gate (2011) 208, 211–12, 222n.118
Integrated Review (2021) 178, 184–9, 193–5, 206–8, 225–6
intermediate-range nuclear forces (INF) 104, 135–8
see also ground-launched cruise missile; Pershing
In-vitro fertilisation see embryology

Joint Intelligence Committee 42, 45, 69
just cause 145–6, 149
just war 128, 145–58, 201–4, 225

Kennan G. F., US Ambassador, BBC Reith Lecture 72
Kennedy J. F., US President 77–81
Kent B., CND General Secretary/Chair 123, 133–4
King George VI 145–150, 158
Kissinger H., US Secretary of State 123
Knox R. 165
Krushchev N., Soviet Premier 79–81

last resort 22, 145, 181, 184
Liddell-Hart B. 41

MacMillan H. Prime Minister 65–9, 73, 77–82, 84, 183, 199
McMahan J. 151–2, 157

Index

McMahon Act (1946) 63
McNamara R., US National Security Advisor 123
Madden C. E., Chief of Naval Staff 26–7
Mason R., MOD Chief Scientific Advisor *see* Duff-Mason Report
Massiter C., MI5 Agent 133
MC-14/3 86
  see also NATO
Meacher M., Secretary of State for Agriculture 170
Media, news 2, 4, 111, 121, 160–1, 164, 167–71, 205, 214–16, 225–6
Medical Research Council 161–4
MI5 132–4
Milch E. 50
Milne G., Chief of the Imperial General Staff 26–7
minimum deterrent 117, 129, 182–6, 192, 197, 207
Ministry of Economic Warfare 39, 47
MISC7 Cabinet sub-committee 98–103, 105, 113, 128–9
MISC17 Cabinet sub-committee 82–3
Molesworth (RAF station) 118, 120
Monte Bello 63
Moscow 133, 136, 152
Moscow Criterion 85, 87, 96–7n.131, 98, 100, 137n.8, 182
Mulley F., Secretary of state for defence 87–8
Multi-Lateral Force (MNF) 82–3
  see also NATO

Nagasaki 55, 178
Nassau 77, 82
National Council for the Abolition of Nuclear Weapons Tests 71
National fire control message (NFCM) 156
NATO 74, 77, 99, 102–12, 127, 131, 135–6, 152–3, 180, 184–5, 190–7, 204
  see also High Level Group; intermediate-range nuclear forces; MC-14/3; Multi-Lateral Force; Nuclear Planning Group
Newall C., Chief of the Air Staff 38–9
Normanbrook (Lord) Brook N., Cabinet Secretary/Chair BBC Board of Governors 9–10
North Atlantic Treaty Organisation *see* NATO
Nott J., Secretary of State for Defence 110, 115–17, 125, 129, 131, 204, 215
Nuclear Non-Proliferation Treaty 110, 113, 192–3, 207–9
Nuclear Planning Group (NPG) 105–6
  see also High Level Group; NATO

Official Secrets Act 12, 59, 74, 133
OGD 80/23 *see* future United Kingdom strategic nuclear deterrent force, The
Owen D. (Lord), Foreign Secretary 86–9, 182, 212, 214, 216

Pacem in Terris 125
Pandemic *see* Coronavirus
Papal Encyclical *see* Pacem in Terris
pastoral 126, 152
  see also Pacem in Terris; American Catholic Bishops
peace camps 118, 133
  see also CND; Faslane; Greenham Common
Penney W. 59, 63
Pershing 105–6, 119, 136
  see also intermediate-range nuclear forces
Polaris 76–7, 82, 84–9, 98–104, 107–20, 126–9, 135–6, 182, 190, 197, 199, 224–5
sales Agreement 77, 83
  see also Chevaline
Pope John XXIII 125, 128

Portal C., Chief of Air Staff
  Strategic bombing strategy 43, 47–8
  Nuclear development 58–9, 63
Poseidon 112
prestige 178, 196, 198–200, 209
Prime Ministerial Directive 156
proportionality 27, 51, 146, 153, 184
Protect and Survive 121–4, 152, 200
  see also civil defence
Pym F., Secretary of State for Defence 99, 102–19

Quinlan M., MOD deputy under-secretary (policy) 83, 104, 109, 127, 150, 181, 214

Reading S., Chair Women's Voluntary Service for Civil Defence 14–5
Reagan R., US President 136
right intent 145–9
risk 167–71, 181–6, 205
Rodgers W., Labour defence spokesman 108, 111, 119
Rodin D. 151–7
Ruddock J., CND Chair 132–3
Russell B., CND activist 74

Sagan C. 123
Strategic Arms Limitation Talks (SALT) 98, 99, 107, 113
Sandys Defence Review (1957) 70–1, 200
Saville inquiry 204
Scowcroft B. 123
Siemes P. T. 61, 91
Sinclair A., Secretary of Air 29, 34, 45–9
Singleton report 44–7
Skybolt 76–7, 135
Snape P., MP 117
Spaight J. 28, 41–3, 48–9, 214
Sputnik 64, 72

SS20 90, 104–5
  see also intermediate-range nuclear forces
Stokes R., MP 46–9
Strategic Defence Review (1998) 170, 183, 205
Strategic Defence and Security Review (2010) 183, 200
Strategic Defence and Security Review (2015) 183, 188, 198, 200
Strath report 12–15, 69–70, 121, 123, 200–1
Submarine-launched ballistic missile (SLBM) 102, 112, 186, 187
Submarine-launched cruise missile (SLCM) 102, 112
Successor
  Polaris successor 86, 88–9, 100–4, 109, 110, 112–15, 118, 128
  Trident successor 179, 183, 189–92, 196–9, 208, 210, 212, 226
  see also Dreadnought
Suez
  Crisis 71, 198
  East of 84
Super Antelope *see* Chevaline
supreme emergency 2–3, 146–7, 153–6, 202–3, 225

tactical nuclear forces 86–7, 99, 105, 109, 125, 152
Teutates 190–1
Threads 123, 213
Tomahawk *see* ground-launched cruise missile
Trenchard H., Chief of Air Staff 23–7, 31, 41, 48, 50, 109, 202
Trend B., Cabinet Secretary 10
Trident 89, 98, 103, 104–20, 127–30, 132–6, 161, 171, 178–86, 190–1, 196–9, 204–12, 224–5
Trident D5 115, 128–30

Index

Valiant 64
Vance C., US Secretary of state 106
Vulcan 64

Walzer M. 150, 153–5
War Game, The 1–15, 70, 117, 123, 131, 182, 200, 214, 216
Warnock M. 4, 162–69, 170, 204–5
Washington Declaration 56
Watkins P. 1–14
 see also War Game, The

Wells H. G. 27–8, 70
Wheldon H. 6, 9–10
When the Wind Blows 124
Whitelaw W., Home Secretary 109, 121–2
Wilson H., Prime Minister 10, 12, 82–5, 135
Women for Life on Earth 120
Wyndham Goldie G. 9, 11, 124, 128

Zeppelin 19–20, 193

EU authorised representative for GPSR:
Easy Access System Europe, Mustamäe tee 50,
10621 Tallinn, Estonia
gpsr.requests@easproject.com

www.ingramcontent.com/pod-product-compliance
Ingram Content Group UK Ltd.
Pitfield, Milton Keynes, MK11 3LW, UK
UKHW021830210426
5322IPUK00004B/120